Understanding Models in Earth and Space Science

Understanding

Models

in Earth and Space Science

Steven W. Gilbert
Shirley Watt Ireton

Understanding Models in Earth and Space Science
NSTA Stock Number: PB174X
05 04 03 4 3 2 1

Library of Congress Cataloging-in-Publication Data
Gilbert, Stephen W.
Understanding models in earth and space science / Stephen W. Gilbert, Shirley Watt Ireton.
p. cm.
Includes bibliographical references.
ISBN 0-87355-226-1
1. Science—Study and teaching (Secondary) 2. Science—Mathematical models. 3. Mathematics—Study and teaching (Secondary) 4. Models and modelmaking. I. Ireton, Shirley Watt. II. Title.
Q181.G395 2003
507'.1273—dc22

2003015972

This publication was developed under Assistance Agreement No. 1W-2905-NAEX awarded by the U.S. Environmental Protection Agency. It has not been formally reviewed by EPA. The views expressed in this document are solely those of NSTA, and EPA does not endorse any products or commercial services mentioned in this publication.

About the cover: The model on the cover is called a *solar system simulator*. It can be set to show the relative positions of the planets at a given point in time. As every model has inherent limitations, it is also the case with this model. Specifically, it is necessary to modify this model to fit scale. The model's instructions state that if we "assume the real Sun were six inches in diameter as it is in the model, the real Pluto would then be the size of a small grain of sand and would be nearly ½ mile from the Sun."

Contents

Introduction

Why Teach With Models?

The National Science Education Standards (NSES) emphasize the use of models in science instruction by making it one of the five unifying concepts of science, applicable to all grade levels. The NSES recommend that models be a focus of instruction: helping students to understand the use of evidence in science, to make and test predictions, to use logic, and to assemble their own understanding of how things work. This book is designed to help Earth science educators to effectively use models as they teach.

Using models is a common form of communication. Ask someone on the street for directions and watch as his or her hands, waving and pointing, model behavior for you to follow to reach your destination. Children on a playground will use sticks, stones, and leaves to sketch out a football play or to represent dishes for an imaginary meal. This is because the simplest type of model is the placeholder—a rock or stick can represent anything in the imagination if the sole task is to establish position. In science, modeling has a similar goal—to mimic in some way the particular behavior of a system and thus communicate a better understanding of this behavior—both to the builder of the model and to other learners.

Model building can help students assemble their seemingly fragmented knowledge about concepts and relationships into larger, more clearly understood constructs. The act of building a model requires students to think about and discuss a science concept, breaking it down into pieces (facts) and considering how (and why) those pieces are related. Choosing a useful model is one of the ways intuition and creativity come into play in science.

As you will read, most models are designed to resemble a real thing—a target—in the world. (The exception is mathematical equations, which may be abstract.). A model can be qualitative, helping students to understand structure or form, a model can be quantitative, helping students to understand function or relationships, or a model can have both qualitative and quantitative aspects.

With this much variety in the formats of models, it is not surprising that models can be misused. A frequent cause of students' science misconceptions is confusion between the characteristics of a model and the characteristics of the real thing it represents. Four questions, used routinely during instruction, can help prevent these misconceptions:

- How does this model work the same as what it represents?
- How does this model work differently from what it represents?
- What are the strengths of this model? The weaknesses?
- How does this model compare and contrast with what it represents?

How Did This Book Come To Be?

After the release of the NSES, the National Science Teachers Association began the work of helping teachers and administrators to use the new tool. NSTA published a series of *Pathways to the Science Standards*, written by educators to help other educators, and made the first goal of its "Building a Presence for Science" program one of providing schools with copies of the Standards and a point of contact to guide their application.

As the Standards gained use and the states adapted them to fit more local needs, teachers began to identify gaps between the Standards' recommendations and their preparation and practice. One of these gaps was the use of models in teaching science. Recognizing this, NSTA sought partners for advice and assistance. As nearly all scientists are builders and users of models, NSTA approached the lead science agencies of the federal government.

The U.S. Environmental Protection Agency's Office of Pollution Prevention and Toxics was eager to use its expertise to help educators learn to use models and to help students learn to use EPA's data and models. Cathy Fehrenbacher, Dick Wormell, and Georgianne McDonald represented the EPA on this project. The EPA is one of the most accomplished builders of models in the world. Data to construct EPA models are collected by the National Aeronautics and Space Administration (NASA), the National Oceanic and Atmospheric Administration (NOAA), United States Department of Agriculture (USDA), United States Geological Survey (USGS), and other federal, state, and local government agencies. EPA's models represent environmental systems, helping citizens to understand impacts upon these systems and helping EPA do its work.

Of EPA's internal policies and parameters for models, four in particular helped guide this project and assisted the thinking and writing of the authors. A first parameter is that models must have transparency: The assumptions made by a model builder must be clear to the user of the model. The implication of this point for educators is that a model being used in instruction should be appropriate and comprehensible to the students.

A second parameter of model building is that a model must be calibrated: It has to correspond to the real target being modeled. Calibration helps a user know the conditions and applications that a model can be used for. The implication of this point for educators is that they should help students know that models are tools to find answers and are part of a process of discovery and learning. As with any tool in science—telescope, pH meter, electronic balance, and others—a model must be carefully tested and adjusted to make sure it gives reasonable and reproducible results.

A third parameter is to *not* use a model if there are real-world data that are acceptable and appropriate to answer the question being asked. There are two implications of this point for educators. First, the availability of real-world (and even real-time) data on the Web provides an authentic way for students to use evidence to ask and answer questions in science. Second, students' collection of their own data in the real world provides a multi-sense learning opportunity with a system that a model can only approximate.

A fourth parameter of model building is that models must be peer reviewed. Modeling, like all science, is a team effort. The ideas of many contribute to the work of a model builder—to focus thinking, suggest untested variables, and provide an unbiased appraisal of how well a model works. The implication of this point for educators is that model building is an excellent opportunity for cooperative learning and for emphasizing the nature of science as a process rather than as a collection of facts.

Goals of This Book

Understanding Models in Earth and Space Science is designed to help the practicing teacher or the preservice teacher recognize and effectively use all types of models with students. The chapters begin with a discussion of the most commonly used (and misused) models: verbal models. The culminating chapter is on using models as a means for true inquiry teaching. Throughout, each chapter is filled with examples and background for Earth science teachers.

A student's logic, creativity, and understanding of science concepts are clearly and authentically displayed when the student correctly builds and uses a model. EPA and NSTA have worked to create this book as a tool to assist teachers in the classroom achieve the goal of teaching students to demonstrate an understanding of the variety of models. We welcome your comments and suggestions on this publication.

Understanding Models

Overview

Models are essential to the learning of science, as they are essential to learning in any field. The *National Science Education Standards* identifies models and model-building as unifying concepts in science. In *Benchmarks for Science Literacy*, the American Association for the Advancement of Science recognizes the use of models as one of the four "common themes" of science—along with systems, constancy and change, and scale. Despite this, teacher preparation courses seldom emphasize understanding and effective teaching with models. Teaching materials generally do not even refer to models in relation to science learning activities, despite the fact that most such activities are concerned with model building.

Chapter 1 defines and discusses models in a broad, and perhaps unusual, way. In particular, the chapter stresses the framework of personal models that underlie science and learning across fields. Subsequent chapters will deal more with particular kinds of expressed models that are important in science and science teaching: physical models, analog models and plans, mathematical models, and computer simulations. Throughout, the book examines how all models are important to science, how they are used, and how to use them effectively. They can and should be used not only to teach science, but also to teach students something about the process of learning and about the nature of knowledge itself.

What Is a Model?

A model is a system of objects or symbols that represents some aspect of another system, called its target. We use models every day in conversation, to learn, to experiment, and to make predictions. Model building is at the heart of learning; for models

to be useful, we must clearly distinguish models from their targets. Otherwise, we can end up with misconceptions about both what we have learned and the process of learning itself.

A model relates to a target through certain features that are similar in both the model and the target. We may call this characteristic a correspondence. The parts may look the same, behave in the same way, or may simply correspond through a relationship that is understood. Although models do not have to look like their targets, learners must understand how they represent the targets. In other words, models may have to be interpreted to make sense. Consider the use of a vacuum cleaner to represent a tornado as an example. To a naïve observer, it may not be clear how the former represents the latter, because there are many possibilities. As a result, the observer may focus on the wrong characteristics and thus arrive at an erroneous interpretation of what the models stand for.

Figure 1.1

The model (vacuum cleaner) represents its target (tornado) in that they both suck up whatever is in their path.

Models and How We Learn

Models are far more common in our daily lives than most of us realize. Certainly we recognize some models because they are so obvious, as in the statement, "The tornado roared through town like a giant vacuum cleaner!" In this sentence, the vacuum cleaner, as a model, is used to represent the tornado, its target. More specifically, the suction of the vacuum cleaner is being compared to the suction of the vortex of the tornado. While most of us can interpret the meaning of the analogy rather quickly, anyone who was familiar with tornados but had never seen a vacuum cleaner before might draw some interesting—and erroneous—conclusions.

To understand why models are fundamentally important to us, we must first examine how they underlie learning in all fields, and in our everyday lives as well. We learn, of course, by taking in information and making sense of it. Many parts of our bodies learn; our immune system, for example, learns to recognize potentially harmful substances. However, as teachers, what we are concerned with is generally called *cognitive learning* and takes place in the brain. It is the way we consciously recognize our world.

We receive information about the world around us through our five senses: touch, smell, sight, taste, and hearing. We have created many sophisticated tools—such as telescopes, thermometers, sonar, and pH meters—that extend and enhance our five senses. Nonetheless, the senses themselves still define the limits of our abilities to receive information about the world.

We also receive information from sense receptors located within our bodies. Sense receptors provide us with information about our location in time and space, our balance, and our general state of being. They pick up information internally and send it to the brain. Like the information from external receptors, this internal information helps us determine our responses to the world. Although most of the work of our sense receptors is not conscious, it is crucial to our ability to operate effectively in our environments.

Another source of information influencing us is our memories. No one knows exactly how memory works, but it is clearly embedded in the chemistry of the nervous system. Our memories include our recollection of events, as well as our stored experiences in processing and responding to information. When we respond to stimuli, we create patterns in our nervous systems. We remember which patterns help us survive and achieve our goals, and which do not. The details of our memories fade, but the patterns of recognition and behavior we establish over time become the guiding elements of our lives.

When the brain integrates information from these three sources—our senses, internal receptors, and memory—the result is a mental construct called a *percept*, the root word for perception. Remember that all of the information received by the brain

is in the form of chemical signals. The signals may come from many sources at any given instant. In fact, when we sense danger or excitement, we may be flooded with signals that heighten our awareness and memories of an event; when we are bored, the level of information we receive from all sources is low. One way or another, the information from our senses, internal receptors, and memory must occur for us to consciously perceive the world.

Sometimes we may respond to information before we can make sense of it, as happens when we hear a strange sound at night or narrowly avoid stepping on a snake. Our response means that our brains can recognize sensory and internal information and assemble it into the elements of action without our conscious minds. When we sleep, our bodies continue to receive information, so we can awaken when we are touched or when someone calls our name. When we are conscious of our surroundings, we may believe that we are seeing the world exactly as it exists, like a projection on a screen: In fact, we are not; rather, we are seeing a *mental model* of the world we have constructed from the information our senses, sense receptors, and memories give us.

Perception and Our Mental Models

The percept is a momentary mental model of the external world. It is a fluid and variable idea that changes and flows from moment to moment. Consider the idea that it is artificial (constructed) rather than a direct unsullied image of the world. Why is this so? First, all sensory inputs must be interpreted to have meaning. Sounds, for example, are manifestations of vibrations in a medium such as air or water. The vibration itself is not sound; rather sound is a mental interpretation of the input from receptors in our bodies. Tastes, smells, sights, and touches are similarly derived from inputs to receptors that convert their energies to electrochemical impulses, in somewhat the same way that a television camera converts light to electrical energy.

Since we only perceive our world in this way, we have no idea if it could be perceived differently. Other species of animals, for example, may interpret these same inputs in different ways. In fact, it is unlikely that all humans perceive the world in exactly the same way. In an unusual phenomenon called *synesthesia*, sensory inputs overlap and combine with interesting results. Other people may be perceived in colors, rhythmic noises may be seen as a series of colored dots or dashes, and letters of the alphabet may be visually perceived in different colors or associated with sounds. People with synesthesia perceive the world in a different way than those without this condition, but who is to say which perceived model of the world is more accurate?

That we construct what we see is also apparent when we understand that the input from our senses is not continuous, but is fragmented. Our senses constantly pick up information from several sources, but it is unusual for a stream of information to

be continuous. Chances are you are reading this page holistically rather than word-by-word and letter-by-letter. You are essentially sampling your environment, allowing your brain to fill in the gaps to provide you with an apparently seamless picture.

Or consider the mental processing you employ during a conversation. You may constantly miss words and inflections, yet you are unlikely to be aware that you are missing them because your brain fills in the gaps so smoothly. When you suddenly recognize you have missed something, you may be able to interpolate the missing information, especially if the context is familiar to you. But when the topic is not familiar, you must be more attentive; you then have less ability to fill in the gaps accurately and are far more likely to become confused and lose the meaning of the conversation. Little in our lives is exact, and probably one of the least exact characteristics of life is our mental model of it.

In the course of living, we constantly generate our mental model; it is in dynamic flux. We interpret the incoming signals according to the capabilities of our mental apparatus, our memories, and our emotions. The intrusion of emotions may well provide an unintended context that can result in a misinterpretation of the conversation. All of these factors make knowing and learning a relativistic and probabilistic affair, for we must depend upon context when we interpret our sensory input. Thus, we have only a limited probability of perceiving the world as it may actually exist.

TELEVISION AND MENTAL MODELS

Television provides a rough analogy for the creation of the mental model. Imagine you are on a totally closed ship in space. You must use cameras to see outside. The cameras receive light from space and transform it into electrical currents. These currents run through a highly sophisticated computer where they are clarified, interpreted, and transformed back into an image on a computer screen inside the ship. The computer fills in gaps by supplying information that is missing from its stores of patterns. Inside the ship, you cannot observe reality directly. All you can see is an image of reality. Your view of reality is subject to disruption and distortion by solar storms that cause static in the equipment or by computer malfunctions. In addition, the computer may interpret its mission and commands differently than intended, and as a result it may interpret images in ways that are inconsistent with the incoming information. Yet you would accept the images produced as reality, because you have no other way of knowing.

Using Mental Models in Science

Philosophers differ about the question of whether there is or is not an objective world—one external to ourselves. Science assumes that there is such a world and that it is predictable. We can refer to that world as an external reality, while we should keep in mind that we possess an internal reality as well. Following from this reasoning, internal reality is every bit as "real" as external reality.

Our internal world is the world of the mental model. Mental model building is the essence of learning and constructivism. Mental models are hierarchical in that we may have a small, real-time mental model (playing now, in our minds) and larger framing models that we maintain and through which we interpret the world. When we receive new information, we have two options: we may discard it as incompatible with our existing mental models, or we may add to or modify our mental models.

If a friend seriously suggested that the Earth is flat, we probably would not use this information to modify our mental model of the Earth, but we might modify our model of the friend. We receive a great deal of information (such as the idea of a flat Earth) that is lost completely because it has no meaning to us—i.e., it does not add value to our mental models to keep it.

On the other hand, if we just found out that a hurricane is blowing up out on the Gulf of Mexico and is threatening our coastal city, we could call on our mental model to know what to do. Information about hurricanes is incorporated into our model of the world and may persist in memory long after this particular hurricane has passed. Later when we survey the damage, we learn about what a storm of this type can do. We store this information away, too. It is then part of our knowledge, which we may define loosely as the sum of our working mental model of the world. All parts of our knowledge structure correspond to elements in the external world. In this sense, then, knowledge is a model of the world.

The human brain cannot and need not process all the points of information it might theoretically be able to pick up from the environment. It needs only enough information to build a model that will enable the body to survive and reproduce.

PARTS OF THE MENTAL MODEL (IMAGES AND PROPOSITIONS)

Because we as teachers deal with learning, ideas, and knowledge, we need to be aware of how mental models are expressed. Underlying our mental models is a model that is embedded in the unknown workings of the brain. This embedded model is our "schema," a term you may recall from your study of

Piaget. Our mental models arise from our schemata. Mental models are symbolic, in that they are constructed of symbols: words, letters, and images and related symbols in the other sense domains. These symbols stand for various components of our world. They are put together to create meaningful mental models. Strings of symbols to which we give meaning are called *propositions.* A sentence is a proposition, as is a mathematical formula or equation, or a musical score. Another major component of mental models is the *image.* Humans both think and express their thoughts through propositions and images, of which images may be the more valuable for understanding and remembering. Our brains will assemble a mental model when we need it, based on the information it has collected. For example, when you say the word "volcano," a host of images and propositions will be drawn together in your mind. A large part of your mental model of a volcano will be the images you have seen either directly or in pictures. The statements you make about the volcano are propositions and constitute the other major element of your mental model of the phenomenon.

Figure 1.2

Mount Mageik volcano, Alaska. Images such as this help create our mental model of a volcano.

Photo courtesy of U.S. Department of Interior

Conceptual Models

Concepts are units of thought. They are categories of characteristics, objects, or relationships that exist—or are held to exist. We generally refer to concepts by labels: words, letters, numbers, signs, and symbols. Concepts are categories because characteristics, objects, or relationships in the world are generally variable. Therefore, a concept must be sorted into a category according to its features and the needs of the person doing the sorting.

Concepts are a kind of formalized mental model. In fact, we create concepts to organize our mental models. However, not all our mental models have a formal conceptual organization; in dreams, for example, we may encounter phenomena that are unusual, to say the least.

Conceptual models are mental constructs consisting of symbols, images, and propositions that together characterize a certain category of events. Consider the category "blizzards." Within this category we can include propositions related to their formation, appearance, and importance, as well as images of them. We can also include any personal experiences (with associated emotions, physical contact, taste, smells, etc.) we have had with blizzards, as well as associations with related phenomena (such as snow showers or ice storms). Taken all together, these propositions, images, and experiential memories define our personal model of blizzards. Others may form their own personal models of blizzards, which may not be exactly the same as ours.

Rich conceptual models evolve from experience and come from exposure to examples and non-examples. The more senses there are involved in creating a conceptual model, the richer and more robust the model will be.

DREAMS

Explanations of dreams have varied across the ages, ranging from visitations by the gods to spontaneous random firings of the brain. Dreams perhaps best exemplify the idea that neural processes manufacture reality within us. Some writers have speculated that dreams are a completely internally generated reality; they are often fantastic because they lack external information to give them form and structure. In some cases, victims of stroke report having hallucinations that seem quite real. Internal reality is highly relativistic, but we have no way of knowing for sure (except through agreement) whose reality is most consistent with the external world.

Expressed Models

Knowledge begins with the information we take in. We constantly know our surroundings. If we wish to communicate with others, we must develop the means to do so. We do this by creating representations of our mental models. These representations are collectively known as *expressed models*. When we express our ideas, we do so with models. All the artifacts we build to communicate with each other are models that represent and express our mental models. In terms of understanding learning, this idea is very important: It is important because we cannot communicate a picture of the external world; rather we can only communicate our ideas about it. To develop an expressed model, we must first take in information from the outside world, translate and process it to create a mental model, and then translate that mental model into a new medium to create the expressed model.

The many ways through which we communicate all entail expressed models. For example, when students write reports, they are creating models. In each report, words represent objects, characteristics, and relationships in the external world, as a student perceives them. A novel is a model of a world created in the author's mind; so is a work of art, a mathematical equation, or a musical phrase. We are familiar with the many objects that are routinely considered to be models, but the models concept actually describes many interrelated constructs.

Consider a diagram of the Earth orbiting the Sun, as shown in Figure 1.3. This diagram is, of course, an expression of a mental model, not the real Earth-Sun system. Above the diagram is the caption "The Earth orbits the Sun." Each word corresponds to or modifies an element in the diagram. Both are expressed models; one is an image and the other is a proposition. The proposition, however, has the diagram as its target, while the diagram has the mental model as its target. The diagram and caption are models of models. Interestingly enough, the greatest part of our knowledge does not come from direct observation, but from studying the models of others.

Figure 1.3.

The Earth orbits the Sun.

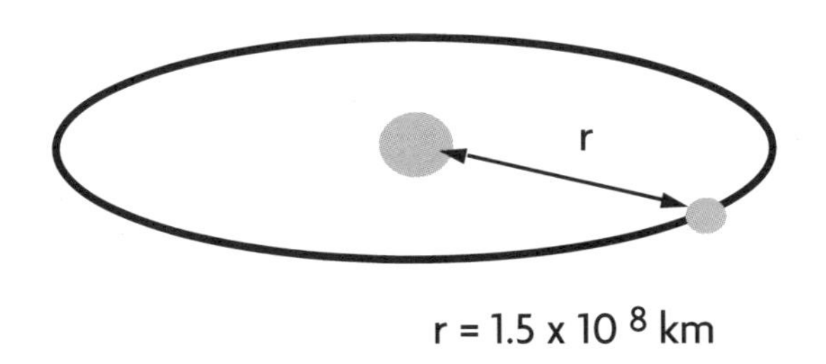

$r = 1.5 \times 10^{8}$ km

Clearly we could continue to construct models of the Earth-Sun system in many media. Which we choose depends upon our purposes. As a general rule however, propositions tend to be weak explanatory models. If we are lost and looking for help in finding a place in a strange town, it would probably be more helpful for someone to draw us a map rather than to simply explain or even to write directions explaining how to get where we want to go. Explanations or written directions are somewhat more difficult to follow than an image. This insight has real applications in the way we teach and how our students learn.

Expressed models are products of our thinking and communication. They are not limited to science, but of course, expressed models in science are constructed to achieve goals that are science-specific. What is truly important in this discussion of models is to understand the nature of our thinking processes. Errors in science, as in all fields, occur because people fail to understand that their opinions and ideas are relativistic and probabilistic models. When we view knowledge in this framework, we can step back and critically examine both it and how it was created. This is a fundamentally different approach to knowledge and learning than when we regard knowledge and learning as attainable in any absolute sense.

It is hard to categorize models because they tend to overlap. In addition, there are as many ways to sort models as there are uses for them. However, we can generally divide them into two groups: those that are concrete and those that are abstract. On a more practical note, as teachers we must be aware that most middle school students deal in the concrete 99 percent of the time. We do not want to load students down with concepts and abstract ideas that are beyond their ability to grasp.

Figure 1.4

A space shuttle scale model

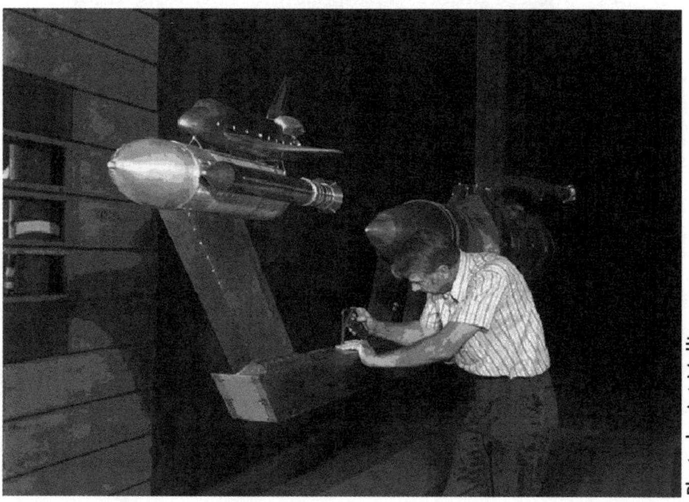

Photo by Art Melliar

Concrete Models

Concrete models are those with which we are most familiar. They are tangible material models that we can generally interpret with relative ease. They are used widely in engineering, science, and science teaching. Concrete models are usually used either to represent the appearance of their targets or to function like their targets. Pictures and pictorials may be considered concrete when they are intended to represent their targets directly. When they become abstract, they may generally be classified with diagrams.

Scale Models

Scale models are intended to look like their targets. Examples of scale models might include such objects as a plastic model of the space shuttle or a plaster model of a particular mountain. Scale models facilitate recognition, but may also be used to identify features of the targets that are of particular interest—for example, a particular volcanic cone. Though scale models have superficial attributes in common with their targets, they seldom share functional attributes. Such models may be specifically related to a particular object—as in the space shuttle model as shown in Figure 1.4—or may represent a whole class of objects, as in a model of an imaginary rocket of the future. Either way, the primary concern is with appearance.

Scale models tend to be static; the functional similarities they may share with their targets tend to be very limited. This is because of the limitations of scale. Some models built to look like their targets are built to full size, of course, but they are essentially scale models in which the scale is 1:1. Engineers and scientists may find it useful to build life-size models to better assess characteristics of the target without the complexity that characterizes a fully functioning prototype.

LIMITATIONS OF SCALE

A scale model has a simple physical limitation: If a body changes in three-dimensional size, its mass changes by the cube, while its surface area changes only by the square. In other words, if a model doubles in linear size in all directions (a factor of 2), its surfaces will increase by four times (2^2), while its mass will grow by a factor of 8 (2^3). Thus, a toy-size scale model of the space shuttle with exactly the same proportions as the original would be extremely flimsy, while the same scale model enlarged to full size would be far too massive to fly.

SCALE MODEL EXAMPLE

To understand the limitations of scale, think about the problems involved in carrying out the following activity with a class. Take students outside to construct a scale model of the solar system. Where there is plenty of space, use balls of proper relative size to represent the planets and sun. Position the students holding the planets (balls) at appropriate distances from the sun and each other. When everyone is in place, point out that while this model is to scale, it does not show motion. What other shortcomings of this model can students recognize?

Functional Concrete Models

Functional concrete models are intended to represent certain functional relationships of their targets. They have relatively less emphasis on retaining scalar relationships and on accuracy of appearance. A classroom model of the solar system is an example of a functional model. It is not intended to represent the solar system to scale—a very difficult thing to do in the classroom. Instead, this model's purpose is to illustrate the relative positions and motions of the Sun, planet(s), and moon(s) in relation to each other.

Figure 1.5

Example of classroom model of the solar system

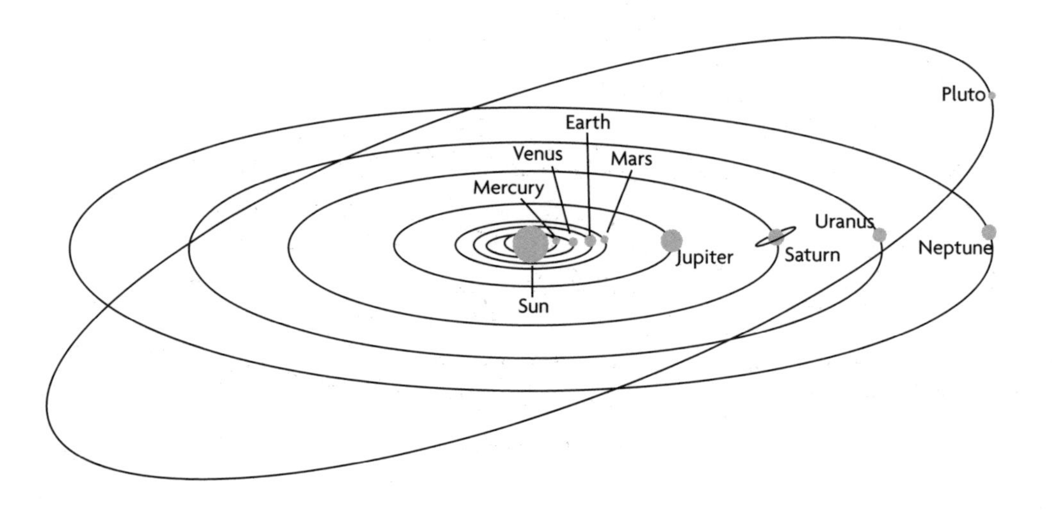

Similarly, pictorial diagrams of the solar system may well distort the relative sizes and positions of the various bodies in the system in order to illustrate their arrangements. These diagrams can be highly misleading when they are used for instruction, since students may not be aware of these distortions. Students, for example, may accept the elongated ellipses often used to illustrate planetary orbits as an accurate representation of what orbits actually look like.

Concrete models are essential tools in science and in teaching science. However, both students and teachers must recognize and acknowledge their shortcomings.

CONCRETE MODELS AND THE LABORATORY

Laboratory experiences are in essence about using and building concrete and functional models. An experiment has little meaning in itself; rather its meaning is the extent to which it models some larger aspect of life and experience. If we drop a cube of colored ice into a pan of warm water, we can observe the flow set up by the difference in densities of the cold and warm water. For this lab to have meaning, however, we have to relate it, as a model, to larger phenomena such as the flow of ocean currents. This functional concrete model has similarities with and differences from this larger phenomenon it is modeling. What is important is that we understand it as a limited model. Science is a way of creating and using such models to build understanding.

Diagrammatic Models

Diagrammatic models include systems diagrams, flow charts, blueprints, concept maps, topographical maps, and similar kinds of two-dimensional line drawings. In these types of models words and symbols represent various objects and relationships, rather than representing a scene directly. .

Graphs are diagrammatic models that show real quantitative relationships; thus they are a combination of mathematical and diagrammatic models. The line of a graph represents real-world changes in one variable relative to another. Values on the vertical and horizontal axes represent real conditions in the environment. For the graph to have meaning, students must understand that there are rules both for constructing and interpreting the graph. Equally important, students must visualize the system that the graph is actually representing.

Mathematical Models

Mathematical models are based on quantitative values and relationships. Some examples include simple mathematical relationships (such as $=$, $<$, $>$), formulas, graphs, and computer models.

Formulas and Equations

Formulas and equations are abstract models that have propositional formats. A useful example of a formula, that which defines our concept of force as a quality of mass in motion, is "force equals mass times acceleration," or, simply: $F = ma$.

This formula expresses a mathematical relationship that exists among properties of mass, space, and time. To use this model we must interpret it and must understand the rules of correspondence. The symbol "=" has a precise meaning, as does the juxtaposition of the "m" and "a," which symbolizes multiplication—itself a complex concept. These operations are real actions that we must take to make the model functional. The abstract model, $F = ma$, does not look like anything we can see. We can see mass and view acceleration, but we cannot witness force. Force exists only through mathematics, and is a good example of a case in which humans have literally constructed reality. Many of the concepts of quantum physics and astrophysics are mathematical, and have no descriptors in the observable world; they exist only in the form of abstract conceptual models.

Equations are mathematical relationships. Though used in science, they reside in the domain of mathematics and obey mathematical rules. Unlike science, mathematical models cannot be validated by observations—they are governed instead by internal logic. An equation may represent relationships between quantities of real-world things, or it may be abstract. In pure mathematics, written formulas may represent mental models—nothing more. In science, mathematical formulae are generally applied to a real-world problem and thus represent relationships between and among real things.

Computer Models

Computer models, among the key tools of science today, are mathematics-based constructs. They are most often used to create images of phenomena, to find and test relationships in complex systems, and to test multiple hypotheses. Computer models include web-based "applets," which provide visual models of the relationship of (usually) two variables; simulations that show visual models of complex phenomena (with or without data entry by the learner); and computer-based tools to manipulate massive databases, which allow modeling of complex "what if?" questions.

Abstract Models

Abstract models use symbols to represent real-world objects, characteristics, and relationships. These kinds of models have to be translated to have meaning.

Verbal-Theoretical Models

All structured writings are verbal models. Words organized into sentences represent mental models that, in turn, represent real entities in the external or internal environment. Some, though not all, such models are theories, of course.

As science teachers, we are familiar with the idea of a theory, although we may not have heard it conceptualized as a model. Following from the discussion thus far, however, the idea of a theoretical model makes sense. As used in the literature, theoretical models are our constructed explanations for what we observe. The scientific theory of the evolution of species, for example, is a model constructed from the factual evidence and the inferences of science.

Theoretical models, unlike many of the models we are used to, are comprised of ideas. The benefit of using the term "theoretical model" in place of "theory" is that it conveys more information about the nature of the idea. It is important to remember that theoretical models are data-based. They are not imaginary creations without a basis in fact; rather, they are based on shared observations, recorded as data.

On the other hand, hypothetical models are proposed causal links or explanations for something we observe or infer. Like theoretical models, hypothetical models develop from preexisting ideas and observations. In fact, hypothetical models are often based on parallel patterns we have observed or know about in relation to similar events. Like theoretical models, they are inferences. If we actually observe a causal link in a particular instance (A causing B) then we have a relationship (in the form of a mental model) that we can communicate directly. But observed relationships are not hypothetical models.

We most often test a hypothetical model by predicting from it. The usual line of reasoning we use when testing a causal hypothetical model such as "A causes B" is this: "If A causes B, then we expect to observe C." "A causes B" is our hypothetical model, while our "If...then..." statement is a prediction.

Idea models are obviously very important in science, as they are in most intellectual activities. When we think about our own ideas as models—rather than as established knowledge in an absolute sense—we gain flexibility in examining our own beliefs and behaviors, a prerequisite of the critical thinking required in good science.

Similes, Analogies, and Metaphors

All models are analogical by nature, but analogies are a particular class of verbal models that require special attention. An *analogy* is a proposition of the form A is to B, as C

is to D, where the A-B and the C-D system have a model-target relationship.

Similes, analogies, and metaphors have played important roles in scientific description and discovery. Science teachers often use them, especially when it is important to make abstract ideas more concrete. However, these verbal models can be particularly misleading, for reasons that will be discussed later.

Characteristics of All Models

Through this text, we will introduce various additional types of models. However, all models, whether mental or expressed, share certain characteristics. These characteristics are important because they underlie both our thinking and the expressions of our thoughts. By their very nature, all models are:

- **Artificial**. All models are human constructs, even if they make use of an existing system or object. For example, if we use an orange to model the roundness of the Earth, we give the roundness of the orange a special meaning, a meaning that it does not normally possess. Because we give meaning to the orange to make it a model, the model is artificial, even if the orange is not. Note, however, that the term "artificial" does not mean "false." It simply means something is created rather than naturally occurring.

- **Utilitarian**. Models are constructed to serve particular purposes. Usually models are not intended to represent all elements of a particular system, but rather to reveal some narrower aspect of it. Information is often deliberately omitted from a model in order to reveal a desired target. For example, a classroom model of the Earth may be useful for geographic relationships, but not useful for geologic processes. We hold ideas and choose models for communication that best suit our purposes, not because they are right in any absolute sense.

- **Simplified**. Models are generally far simpler and contain less information than their targets. Because of this simplification, a model lacks the effects of variables—or attributes—that may be present in its target. The best models do their job with little interference from irrelevant or conflicting features. When we construct mental models, our models usually do not contain all possibilities—only those that serve our needs and purposes.

- **Interpreted**. All models must be understood and interpreted on their own terms. Some require more interpretation than others. A scale model is generally pretty clear on its face, but a road map, for example, requires the user to consult the map's key to make sense of scale, road types, town sizes, and so forth.

- **Imperfect**. Models should never be considered perfect or complete representa-

tions of their targets. Only the target can be perfect. Models are right or wrong only in relation to criteria that define their "goodness of fit." Models imperfectly and probabilistically represent their targets. There are always errors in relation to their fit.

When we construct models, we can determine their usefulness by certain criteria, not the least of which is "goodness of fit" to the target for our purposes. The fit of a model is assessed according to a number of factors, including its

- **Relatedness** to other models, especially other models of the same target. How well does it fit with other models? We are likely to reject or at least hold in abeyance a model that is too different from accepted models. Another term for relatedness is consistency.

- **Transparency**, a measure of how obviously the model fits the target. Some models are very transparent or obvious, while others are rather opaque. When a model is opaque, we don't really get a good sense of what is being represented. Consider the metaphor inherent in using the term "string" to describe the smallest entity in the universe. Because the concept is mathematical, the use of this physical metaphor is probably opaque to many people.

- **Robustness**, a measure of how insensitive the model is to changes in assumptions. In general, the more assumptions needed to link the model to the target, the greater the likelihood that the model will need to be changed or discarded later. The model is robust if it requires few assumptions to be understood.

- **Fertility**, a measure of how much the model explains. The best models explain more about their targets than lesser models targeting the same system. The most fertile models give rise to new understanding and broad insight.

- **Ease of enrichment**, a measure of how easy it is to add to and extend the model. All other things being equal, better models are easier to extend and enrich. We can add to and extend the best models with relative ease.

Models are tools for understanding and are judged by how well they do their jobs. Because models are necessarily simplified, they will not contain all the variables present in their targets, and thus contain considerably less information than their targets. It is critical, therefore, never to confuse a model with its target.

Models need to be accepted or rejected on their own terms as models. When we construct and use models, we should continually critique their usefulness for achieving our goals. This practice should become a routine part of every science class. The ability to describe the differences and similarities between a target being studied and its model is a remarkably useful assessment measure.

Role of Multiple Models in Teaching and Learning

With the background we have just developed, let's examine the value of using multiple models when teaching science. Consider three models commonly used in teaching Earth science to develop a conceptual model of tornados. The first model is a "tornado in a bottle," the second is a two-dimensional drawing of a tornado in cross-section with its parts labeled, and the third is a video of a tornado roaring through a city in Texas. Each model targets specific elements of a tornado and each makes a particular contribution to students' evolving mental models of tornadoes. All are dramatic simplifications of the real thing.

The "tornado in a bottle" (Figure 1.6) is a physical analogue model in which the whirlpool motion of the water in the bottle corresponds directly to the motion of the winds in a tornado. The model fits moderately well, but only in representing one narrow aspect of the tornado. It is poor in other ways: Most of the dynamics of the fluid in the bottle are not the same as those of winds in a tornado; driving forces are dissimilar; and in most other ways, the model is not at all like the target. This is not a *rich* model. That is, it does not have many parts and relationships corresponding to those in the target.

The diagrammatic model of the tornado (Figure 1.7) has more corresponding

Figure 1.6

Tornado in a bottle

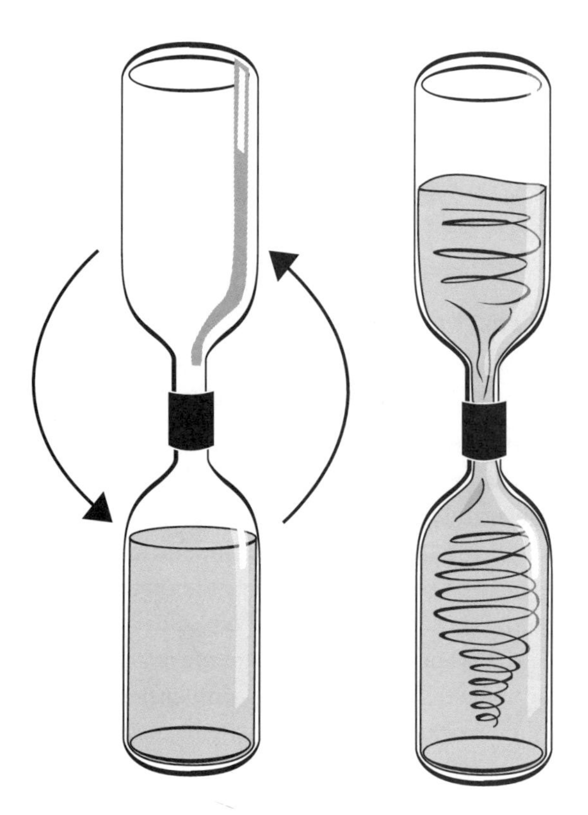

relationships and is intended for illustration. It is not a dynamic model in that nothing moves. It allows students to see the anatomy of a tornado as a static two-dimensional drawing and to label the major parts of a tornado in their mental models. The combination of both the "tornado in a bottle" and the drawing helps expand students' mental/conceptual models in ways that neither model can do alone. Add to this a model that shows the overwhelming destructive power of a tornado—the video—and students should be able to construct a relatively good mental model of the tornado. Certainly students' resulting mental models will be richer than the result from one or two of the models presented alone.

Figure 1.7

Two-dimensional drawing of a tornado

From this discussion, it follows that for students to construct a rich mental model of any phenomenon, they need exposure to multiple models that show different characteristics of the phenomenon. In fact, the richness of students' mental models will be directly related to the number and types of different models of the same phenomenon they explore. A single model is seldom able to impart the kind of understanding that is needed to truly understand a given target. As it is, too many teachers rely only on the textbook. Used alone, the textbook study of tornados will result in a very restricted mental model.

This need for multiple models is true in any field of study and is not strictly related to science education. However, for us as science teachers, it is especially true if we ask students to learn how to do research, or to explain interactions among science, technology, and society or among the cultural dimensions of science. Science is simply another way of learning, and the models scientists create have particular characteristics and purposes.

Conclusion

The National Science Education Standards recommend the use of models as an organizing framework for teaching and understanding science. Not only do models and model building provide multiple frameworks for understanding science, but they also help students make sense of the world. The goal of developing scientific literacy has appeared repeatedly in various national standards: The basis for meeting that goal is an understanding of learning and knowledge in general. In keeping with these principles, teachers of science may take the following actions:

- Present complex concepts through multiple models that address different dimensions of the same concept. The first step in doing this is to identify which aspects of the model need to be understood, and how to convey that understanding. Density, for example, can be understood qualitatively or quantitatively, and can be visually and experimentally presented in a number of different ways. Each alternative model in which students actively engage changes and enriches their mental models.

- Develop students' awareness of the way they think, learn, and communicate through internal and expressed models. The implications are significant, not only for the way students regard their own learning, but also for the way they regard their own knowledge. Certainly there is a difference between thinking that what you know is right in an absolute sense and regarding what you know as one of many possible permutations of models of the world.

- Engage students in actively building and analyzing expressed models of the systems they study. Students should be able to clearly identify targets their models

represent. They should analyze how well the models fit their targets. This involves describing how the model does and does not represent the target, assessing how rich or poor the model is, and perhaps identifying how the model could be enriched in relation to the target. It also includes generalizing to targets beyond that which is most immediate, an activity that puts students into the role of scientific model builders and more clearly identifies the nature of the products of their work. Along with doing experiments, they are seeking to create models of the phenomena they are studying.

- Guide students in analyzing scientific models and testing of competing models against one another. This is particularly true when the models are rival potential explanations for a phenomenon. It is also true when models arise from different knowledge sources, as, for example, science and religion. By analyzing the strengths and weaknesses of the models, students can arrive at a reasoned judgment about the fit of each model to the purposes each is intended to serve. This approach tends to distinguish basic assumptions of different model-building traditions. Analysis also helps students understand how conceptual models can be distorted by different assumptions and purposes and by different ways of accepting and processing information to explain a phenomenon. In addition, analyzing and testing competing models also helps students understand why different individuals might develop different models to explain the same phenomenon.

- Involve students in systematically refining models to increase their accuracy. Students should understand how such refinement could be valuable to both the practice of professional science and to them in their own personal science. Perhaps one of the most valuable potential effects of models-based teaching and learning has little to do with science at all; rather it concerns itself with students' self-awareness—the knowledge that we are unlikely to know everything there is to know about the world we inhabit.

Similes, Analogies, and Metaphors

Overview

Analogy is the foundation for all models, but in this chapter we will discuss analogies as a class of models. Because analogies underlie all similes and metaphors, we will use the term "analogy" either as a specific expression of relationships or as a broad term that includes similes and metaphors. Recognizing that analogies are common verbal models and are also common in science, we'll examine them as modes of expression students will need to understand in order to be literate in science.

ANALOGIES IN SCIENCE

Roy Dreistadt has identified many uses of analogy and metaphor by such noted scientists as Poincaire, Newton, Maxwell, Bohr, Einstein, Darwin, and Wallace, among others. Lord Kelvin wrote that he never really understood a thing until he had made a model of it, and Oppenheimer viewed analogy as an indispensable and inevitable tool for scientific progress.

Similes

Similes are the most commonly used verbal analogical models. A *simile* is a statement in which one system is clearly held to be *like* or similar to another. It is generically defined by the statement "A is like C." For example, the propositions "sedimentary rock is like ice created from compressed snow" or "a tornado is like a whirlpool" are similes. Similes imply but do not directly express the relationship that unites the model and the target.

In the first simile, the ice is a model for sedimentary rock. They are alike because both are formed by compression of much smaller bits of matter—sediment and snowflakes, in this case. In the second simile, the whirlpool is used as a model for a tornado. The simile does not make it clear how they are alike; the reader must infer the target relationships.

Similes are analogies in which part of the analogy is unspoken, but presumably is understood by the reader. This characteristic makes them attractive to fiction writers, in whose art ambiguity may be valued, but sometime creates problems in science, where clarity of meaning is important.

Analogies

The proposition $A : B :: C : D$, which we read as "A is to B as C is to D," formally expresses analogy. The letters in this expression represent nodes such that A corresponds to B; C corresponds to D; and the relationship between A and B corresponds to the relationship between C and D. These relationships are shown diagrammatically in Figure 2.1.

Figure 2.1

Diagrammatic representation of an analogy

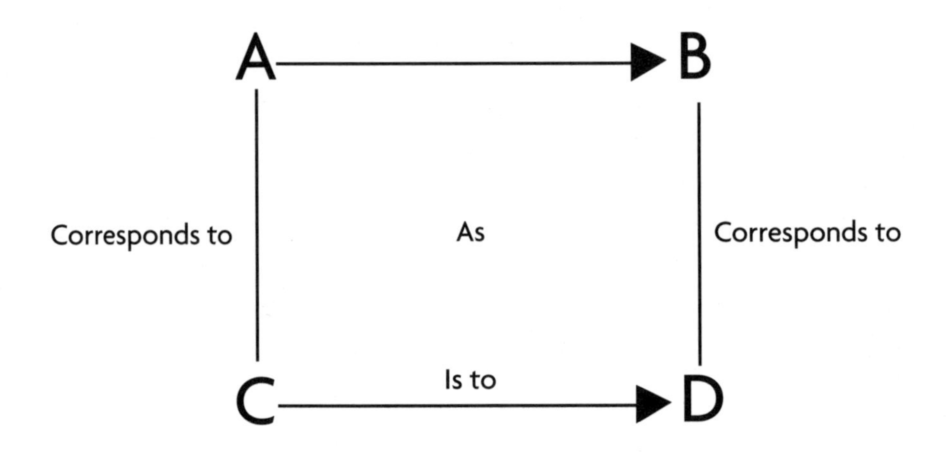

Formal analogies are seldom found in textbooks or classroom discourse, but we can see that they underlie similes, which we have discussed, and metaphors. Let's look at the similes identified previously and write them out as analogies.

- "Sedimentary rock is like ice created from compressed snow" is a simile based on the analogy "sediment : sedimentary rock :: snow : compression ice."
- "A tornado is like a whirlpool" is a simile based on the analogy "tornado : air :: whirlpool : water."

As in similes, an analogy doesn't clearly identify the targeted relationship. For the model to effectively represent the target, the receiver of the analogy has to understand the characteristics of the model and know what relationship is being implied; or, conversely, these characteristics have to be made clear by a full narrative explanation of the analogy.

In the case of the rock-ice analogy, the relationship is fusion through pressure. Snow is fused by pressure into ice without any change in its chemical composition, just as sediment is fused into sedimentary rock. This is a good analogy because the processes are fairly similar, though the underlying chemistry may be different. An analogy fails when the individual receiving the analogy does not know the model, or does not understand which relationship between the model and target is being established. For example, if the individual has never seen snow, he or she might not understand the relationship, and the analogy would not aid in understanding the target.

In the case of the tornado-whirlpool analogy, the corresponding relationship is the vortex. This again is a good and rather transparent analogy. There are a number of parallels to support it, especially if the whirlpool is in a flowing river rather than a bottle. Both the tornado and the whirlpool require sustaining energy and both are created by current differentials. Each can form and disappear as it moves and as conditions change. Materials caught in their vortices, however, may behave very differently, and their effects upon objects in their paths may be very different (or similar, also, which is why analogies often result in new ideas for testing).

Metaphors

Metaphors convey the least information of the three verbal models and are the most likely to be misinterpreted by learners. A metaphor creates an identity between A and C saying, in effect, that A *is* C. Like similes, metaphors are abbreviated analogies.

The simile "a tornado is like an atmospheric whirlpool," for example, translates into the metaphor "a tornado is an atmospheric whirlpool." An identity is created that can be very misleading, since to say something is something else is to imply that all of the relationships in the model will be found in the target.

Figure 2.2

Some metaphors commonly used in Earth and space science

Jet Stream	Cold Front	Shield Volcano	Black Hole	Fault
Magma Cell	Intrusive Rock	Kettle Hole	Glacial Calving	Horn
Soil Horizon	Meteor Shower	Oxbow	Sea Stack	Volcanic Neck
Alluvial Fan	Solar Wind	Aurora	Calderas	Rock Flour
Cloud Seeding	Desert Pavement	Dust Bowl	Up-warped mountains	Flood Basalts

Usually, metaphors are not directly used in science—except as a source for terms to describe new phenomena. The term "black hole," for example, is one of many metaphors that can be found in Earth science. Black holes are remnants of collapsed stars. They are so dense that nothing—including light—passing within a certain distance of them (their event horizons—another metaphor) can escape their gravity. Therefore, nothing can be seen. Black holes are like interstellar holes. This action of a black hole parallels "holes" into which matter and light disappear in much the same way as water disappears into the drain hole in a sink. We can see the relationship among analogies, similes, and metaphors through an analysis of black holes:

- **Analogy:** The collapsed star appears like a black hole and pulls in matter and light in the same way as the hole in a sink drains water.
- **Simile:** The collapsed star is like a black drain hole in space.
- **Metaphor:** The collapsed star is a black hole.

As models, analogies are written or spoken for particular purposes and must be targeted to the audience. This means that the correspondence between node and relationship must be apparent; the target relationships must be essentially the same, or at least similar enough to serve the purpose for which the analogy was intended. In the metaphor of the black hole, the drain hole is a model for aspects of the less well known collapsed star. Drain holes are familiar, and are useful models for conveying the property of the black hole as a "sink" for matter and light.

The original creators of the metaphor may not have been thinking about a drain hole at all. In fact, it is possible that the scientists who gave their discovery the name "black hole" may not have had anything in mind other than an image of a black lightless hole in space. Such an analogy would not have had the richness of a drain hole model, however. Whether or not we know their thinking, it is useful to recreate the metaphor

to make it effective for our own use.

As is often the case, this metaphor has the strong potential to mislead students. A true hole is an absence of the matter surrounding it, while a black hole is comprised of super-dense matter. Drain holes allow water to pass into pipes, while it is not clear what happens to matter entering the black hole; presumably, it accumulates there. Both act as sinks in a general sense of the word, and both appear to be dark, but to our knowledge, they share no other common attributes.

Figure 2.3

Analysis for the black hole analogy

Similarities	Differences	Usefulness and Validity
In the same manner that water moves into and down the drain hole in a sink, matter and energy move toward a black hole. In both cases the motive force is gravity and in both cases movement of materials is one-way.	A drain hole is a space, not solid, while a black hole is a super-dense object. The water in the sink is conducted away by pipes while the matter and energy entering the black hole probably just become part of it.	This is a good metaphor since the similarities are so direct and the model is easily understood. The essential differences can be easily explained and remembered.
Important Relationship: Both the hole in the sink and the black hole act as gravity sinks to remove something from their environments.		

LITERAL SIMILARITIES

In a literal similarity, a known system in nature is usually extended to explain a similar system. This is the basis for generalizing a phenomenon to new examples. For example, the proposition: "sedimentation in a lake is like sedimentation in the ocean" is a literal similarity with the structure "sedimentation : lake :: sedimentation : ocean". Similarly, convection of air over a hot plate may be used to model convection of air over a plowed field in summer, but the relationship is not analogous. Literal similarities are useful and can lead to new insights, but they are not analogies in a true sense. As a matter of semantics, it is useful to distinguish literal similarities from analogies when teaching.

Analogies in Science

Analogies have many uses in science. They

- make abstract phenomena concrete;
- provide new insights;
- suggest directions for research;
- generalize relationships from specific findings;
- develop or extend theories; and
- organize complex multiple phenomena into conceptually related systems.

Generally, scientists use analogies to create recognition and understanding of systems that are new, not readily observed, or not easily understood, by referring to those that are better known. Analogies may simply be descriptive, and hence relatively superficial, as in the case of "oxbow" lakes, in which the only similarity is the similar appearance of the lakes and the bow of an ox yoke. As far as recognition goes, such analogies can be very valuable and minimally misleading (provided, in this example, that the receiver knows what the bow of an ox yoke looks like).

But in science, analogies are most beneficial when they suggest explanations for a newly observed phenomenon. Einstein understood relativity through a thought experiment in which he figuratively rode a light wave. Early explanations of electrical current were visualized through parallels with the flow of water in pipes. The planetary nature of the atom evolved from parallels to the solar system. The list of such useful—if ultimately discredited—analogies is very long.

Analogies can lead to solutions that occur in a flash of insight. Kekule conceptualized the ring form of the benzene molecule in such a burst of insight, when he had a dream of a snake grasping its tail in its mouth. Newton's experience with the apple is another such moment of insight. Aristotle wrote in the *Poetics* that metaphor is a sign of genius. Whether or not this is true, it certainly is important to the advancement of knowledge in many fields, including science.

Because they involve pattern recognition, analogies may help us solve problems by reducing the imagery we need to arrive at a solution. Much of what we learn, formally or informally, is intelligible only because we can find analogous or literally similar patterns. This characteristic is also true in our own social development. Behaviors that work in one situation may be tested in similar new situations. An approach that solves one problem may solve a similar problem in a different context. Explanations that are satisfying in one situation may explain similar phenomena.

ROLE OF ANALOGY

Dreistadt explains the role of analogy this way: "Insight [occurs] when one finds a stimulus pattern (the analogy) in which parts of the form or structure are like the structure of the problem situation, and the rest of the structure of this stimulus pattern (the analogy) indicates how to organize the uninterested materials of the problem or how to reorganize the problem by putting the parts that are out of place into their current place, or both, thereby completing the whole which is then the solution to the problem."

ANALOGY IN THE HISTORY OF SCIENTIFIC THINKING

While analogy is an effective tool for thinking about modern science, scientists today do not consider it by itself to be a stable framework for knowledge. This has not always been the case. Early Western science, especially in classical Greece, was grounded in culturally determined beliefs that the natural world was essentially analogous by nature. Aristotle in particular used craft analogies to explain the processes of nature. His beliefs provided the theoretical basis for his explanations of empirically derived facts, and were bound to other beliefs of the day—especially religious beliefs. These same beliefs continued to influence Western science during the medieval period. One of Sir Francis Bacon's major accomplishments was to begin the Western tradition of separating religion, which tends to be heavily grounded in analogy, from science. This separation enabled the development of a new theoretical basis for science that did not depend upon analogy as a framework for understanding the world.

While analogies play major roles in creative thinking and communication, they are also major sources of misconceptions. Misconceptions especially occur when learners think a model represents all aspects of the target, accepting the model literally. Misguided or naïve acceptance of ideas based on analogy instead of evidence and empirical proof has retarded the advancement of science throughout history. How-

ever, the misuse and abuse of analogies is not a good argument for avoiding them. In fact, we can't avoid them, for analogy is how we learn and think about our world.

Analogies and the Teaching of Science

Analogies are used widely in the Earth and space sciences for description and explanation. As is true with the concept of the black hole, problems of scale, distance, and accessibility often require us to use analogies to visualize key concepts.

Explorations in science are typically guided by analogies. For example, the methods we use to look for life elsewhere in the universe are determined by analogous and literally similar parallels with what we know exists on Earth. Our expectations about life forms on other worlds are fueled by the discovery of life in habitats on Earth that model possible habitats on these worlds.

Our recognition of the meaning of observed phenomena is often suggested by our mental comparisons with systems that are more accessible or better known to us. Early natural philosophers might well have related the folding of rock layers in mountains (a metaphor) to the folding of layered fabrics or blankets. Such an analogy suggests an explanation for mountain folding requiring motion of the Earth and pressure; this explanation would ultimately lead to ideas about continental drift (another metaphor) and then to plate tectonics.

When we teach science, we may look for analogies to explain phenomena, but we must always remember that analogies suggest, but do not prove, explanations. Explanatory models in science must consistently predict expected outcomes, at least within a given range of probability. In this sense, we must be sure that students notice the ways analogies can be misleading if they are accepted uncritically.

As teachers, we should look for analogies in science textbooks. We should also encourage our students to look for analogies, and to identify the relationship the analogy is intended to express. Each time we find analogies, we may emphasize that analogies are models that only represent their targets in narrowly defined ways. We can point out that there are usually many differences between the model and target. Metaphors in particular can pose a problem, since they imply an identity rather than a likeness. Significant learning problems can occur if students do not actively recognize the difference between model and target, and instead accept dissimilar elements of the model as being present in the target.

Figure 2.4

Examples of analogies

...because the trade winds are weak along the equator, some of the piled-up water flows back "downhill" (eastward) creating the countercurrent. (p. 225)
[The creation of an impact crater]...is analogous to the splash that occurs when a rock is dropped in water. (p. 385)
These folds do not continue forever; their ends die out much like the wrinkles in cloth do. (p. 161)
Apparently these quiet prominences are condensations of coronal materials, which are gracefully "sliding down" the lines of magnetic force.... (p. 402)
Some folds...are believed to be the result of compressional forces...that squeeze the entrapped rock much like a giant vice. (p. 161)

Source: *Earth Science*, Third Edition, Tarbuck and Lutgens

We must watch out for two specific pitfalls as we use analogies in teaching science. First, analogies are useful only if the learner understands the comparison being made. If the model in the analogy is not familiar, students will have difficulty understanding the two systems and the relationship between them, which increases the learning burden, which leads to frustration and discouragement. Some text publishers avoid analogies because of the challenge of relating analogies to the lives of a broad and diverse audience of students. To counter this pitfall, we need to know the backgrounds of our students so we can pick appropriate verbal models. In addition, we can give students opportunities to create and share analogies, similes, and metaphors, which may appeal to some students who traditionally have not found science interesting.

Second, when an analogy doesn't work for learners, it is often because they haven't critically analyzed it (which is true in literature as well as science). Rather than telling students what properties a given model represents, we should lead students through a process of critical inquiry to find the important relationships. This process will also help us—and our students—recognize potential misconceptions that are possible from the analogy.

Changing Misconceptions of Analogies

One common misapplication or misinterpretation of analogy is the explanation for the cause of the seasons. Naïve learners may say that changes in the distance of the Earth to the Sun cause seasonal changes.

The source of this misconception is analogous in nature. When we get closer to a warm object, we feel warmer. When we get further away, we feel cooler. This conceptual relationship is so deeply ingrained in our mental models that it transfers readily to an explanation of the seasons. We can reasonably assume that most educated individuals have been exposed to the scientifically acceptable reason for the seasons: that seasons occur due to the tilt of Earth's axis and the revolution of Earth around the Sun. That this idea does not always take hold may be attributed to the powerful influence of analogous thinking.

Where a misconception of this kind occurs, the first step toward changing the model is recognizing the analogy. In the case of the changing of the seasons, we may examine analogous systems, like our experiences with stoves or lamps. Now we can establish a conflict: in the Northern Hemisphere, the winter Sun is closer than in the summer. With the analogous basis for the model clear, the way is paved to introduce the alternative model involving the tilt of Earth. In addition to remedying the original misconception, students may also learn something about how misconceptions are created.

The important point to be learned from misleading similarities is that recognition and analysis of the analogical model—in this case a model causing a significant misconception—can aid in its removal as a barrier to learning.

Engaging Students in Creating Analogies

If the ability to create analogies is a sign of genius, then we should create opportunities for students to exercise that ability. Such opportunities are one way to address the needs of creative students and have the additional benefit of helping all students identify the essential characteristics of the targets. Students with a literary bent might find these activities to be particularly motivating.

The easiest way to create analogies is to begin by creating similes. To do this, provide a stem: "X is like…." Invite students to respond, but also make it clear that they will have to identify the essential relationship(s) they are trying to convey through the simile.

For example, the stem "The Sun is like..." could result in a number of responses: a furnace (both are hot); a heat lamp (same reason); a traveler (moves from horizon to horizon); a visitor (appears periodically); a pivot (anchors the planets gravitationally); a parent (gave rise to the planets); and so forth. Each valid analogy promotes functional insight, in an interesting and entertaining way.

As students create analogies, they will identify literal similarities as well as functional ones. For the stem "A lake is like…" their responses will probably include such things as reservoir, puddle, and pond. Ponds and lakes are to a large degree so similar in form as to be literal similarities. These are not analogies, but do provide an oppor-

tunity to point out and analyze parallels the students themselves construct through this inquiry. True analogous models for the lake might include such things as storehouse, mirror, home, hot tub, recreation center, parking lot, roadway, and so forth. In the act of creating these analogies, students are compelled to recognize the great number of different functions and characteristics of a lake, and they can have fun justifying their creative insights.

A related exercise is to provide students with an analogy that may bring them to view the target in a new way. For example, the question "How is a productive lake like a living creature?" calls upon students to exercise imagination, constrained by empirical fact. Students must first analyze the characteristics of a living creature, and then identify which characteristics apply to the lake. For example, like a productive lake, a living creature consists of interacting systems. Both can be healthy or failing, can engage in constant interchange with their environments, require maintenance of a constant energy balance, and so on. Another way to stimulate thinking on this question is to use the technique of posing the opposite: "What do we mean by a dead lake?"

Avoiding Problems of Teleology and Anthropomorphism

To avoid problems of teleology (giving goal-seeking behavior to an unintelligent system) and anthropomorphism (giving nonhuman beings human characteristics), remind students that models are not the same as their targets. The phrase "water seeks its own level" is a teleological expression in which water is given goal-seeking behavior. While water may return to its own level, it does not seek—a term that implies purposeful behavior. It isn't difficult to see that teleological and anthropomorphic explanations are grounded in analogy.

Some cultures are firmly teleological and anthropomorphic, holding that both living and nonliving things possess anima, or souls. Classical Greeks and medieval Europeans held such beliefs, and the pre-scientific practice of alchemy was based upon the idea that metals had spirits. The transfer of the spirit of gold to base metals, which would turn the base metals into gold, was the much-sought-after prize of the alchemists. Beliefs such as these still exist in various cultures of the world.

Analogies' Learning Goals

Perhaps the best way to use analogies, similes, and metaphors is to address them directly—to teach students something about them. To make maximum use of analogies as models, students should

- Recognize an analogy, simile, or metaphor in science.
- Identify the key elements of the analogy.

- Understand the key similarities between the model and the target.
- Understand the key differences between the model and the target.
- Identify the important parallel relationship(s) linking the two systems.
- Assess whether and how the analogy is valid and useful.

Activities to reach these goals add interest to science, and help students understand the concept of analogies and the science concept itself. As part of the process of analyzing analogies, students process information about both systems, but especially the target system, which is of primary concern to our purposes as teachers.

Obviously, some analogies are rich and others are poor, and analysis should help students determine which is which. One useful technique for analyzing analogies is to have students work in small groups to brainstorm their ideas about similarities and differences, then share their results with the larger class. Another good small group activity is to have students complete a matrix of an analogy. The matrix shown in Figure 2.5 suggests a framework students can use when analyzing important analogies. Note that the order of the steps in the left-hand column does not necessarily correspond to the order of the steps students might use in analyzing an analogy. The order isn't critical, but it should be logical.

Figure 2.5

Steps for analyzing an analogy

Steps in the Analogy	Steps to Complete the Matrix
1. Introduce the target concept	1. Create the analogy and make it explicit
2. Recall the analog concept	2. Map the similarities of model and target
3. Identify similar features of concepts	3. Identify the essential differences
4. Map similar features	4. Identify the crucial relationship that makes the analogy useful.
5. Draw conclusions about concepts	5. Evaluate the utility and validity of the analogy
6. Indicate where the analogy breaks down.	

You may want to have students use the steps listed in Figure 2.5 to analyze the analogy that relates an impact crater to the splash that occurs when a rock is dropped in water. (This analogy is quoted in Figure 2.4.) The relationship of interest is that solid debris is thrown up by the impact of the meteorite in much the same manner as the crown of water droplets is thrown up by the impact of the rock in water.

What are the similarities between the impact crater and the splash? The splash results in a temporary crater in the water caused by the force of the impact. Water previously occupying this space is either moved downward or thrown outward. Similar processes occur when, for example, a meteorite strikes the surface of the Moon.

What are the differences between the impact crater and the splash? They differ in medium—the impact crater exists after the event that creates it, while in water, the crater is obliterated by inflow. They differ in chemical and physical effects—in the impact crater, high pressure and heat are present that may actually change the composition of the rock and soil, both on the Moon's surface and of the asteroid itself, which may be obliterated. In a water splash, the rock continues on and is not altered by the impact.

We can conclude that there is validity and value to the splash analogy, but it is limited. To pose a problem for students, consider asking them to propose alternative analogies for the same phenomenon. For example, are there other properties of an impact crater (disintegration and blast) that might make a bomb a more appropriate model?

This brief analysis demonstrates the potential that analyzing analogies has for stimulating critical thinking. Students need to be mentally engaged with the concept of impact crater in order to complete such an analysis. The information students need to complete this kind of analysis is generally available in a standard textbook, and the analysis calls attention to this information.

Once students have conducted several formal analyses of analogies, continue to have them identify and discuss analogies and metaphors when they (or you) find them in the text. Simply asking, "What is the key relationship in this analogy?" will raise students' awareness that these models are worthy of special attention and can trip them up if they aren't careful.

Conclusion

Science evolves through the development of theoretical models that are consistent with our empirical observations. Theoretical models provide explanation and lead to subsequent exploration. In modern science, analogous models inform theory, but they do not by themselves constitute theory. In contrast to the science of antiquity in which theories of the world rested largely on reason and analogy, modern science demands that the models it creates be logically supported by empirical data and evidence.

Despite the potential for misinterpretation that exists when analogies are used, science cannot proceed effectively without analogous models. Because such models are critical to thinking in science, students should be able to understand them and their limitations, and employ them with the same caution that one should exhibit when using any model to understand a target. And because analogies are a wellspring of creativity, students should—from time to time—be asked to brainstorm and justify analogies for important systems. In this way students can experience the joy of creativity and the pleasure of critical inquiry while learning the concepts of science.

CHAPTER 3

Concrete Models

Overview

For most of us, the term "model" conjures up images of plastic airplanes or toy racecars. Concrete models are those that we can manipulate physically, as opposed to abstract models, which we manipulate mentally (or on computers, which are extensions of our mental processes). Most concrete models lie somewhere on a continuum between those that are perceptually similar to their targets but have few or no functions like them, and those that are functionally similar but do not look like their targets. Scale models are representative of models in the first category, while those in the second category are concrete analogue models.

Some concepts in the Earth and space sciences—such as plate tectonics, groundwater migration, cloud formation, and stream flow—are difficult to teach directly. Concrete models allow us to portray aspects of these concepts in a hands-on manner. For example, we may create a cloud in a terrarium by placing hot water in the bottom and pans filled with ice across the top. In this instance, we are creating a concrete analogue model. The terrarium system functions, but does not look like, the natural system that it targets. Instead, the ice pans are analogous to cold air layers, and the hot water in the bottom is analogous to water sources on the ground, such as surface ground water, transpiration, and lakes.

Table 3.1

A table of suggested concrete models

Earth science concept	Common physical model materials
Plate tectonics	Paper puzzle of tectonic plates (geo-patterns) Convection currents of water and food coloring Melted wax crayons Silly Putty™ Shake table
Groundwater migration	Clear plastic column filled with types of earth Unglazed flowerpot PVC pipe or open-ended cans pounded into ground
Water cycle	Cloud formation in a bottle Terrarium
Stream flow	Stream table or trough Suspensions of salt, clay, or fine sand in water Soil-filled paint tray
Structure of the Earth	Hard-boiled egg Clay or plaster of Paris volcano Tennis ball Classroom globe

(Note: Illustrations of above models can be found in *NSTA's Project Earth Science* series.)

Scale Models

Scale models represent a class of models intended to look like their targets. They are larger or smaller than their targets, but retain the proportions and defining characteristics that make them visually similar. While we are most familiar with visual models, analogous models also exist in relation to the other senses (hearing, taste, and so forth). For example, a duck call is an auditory model of the call of the real duck, and aspartame models the taste of sucrose, or cane sugar. However, most models we deal with in science are visual.

Scale models, while limited in their function, are used to facilitate recognition of the target. The scale doesn't have to be exact, as long as there is enough proportional similarity to do the job for which the model is intended. For example, a generic model of a shield volcano may not directly resemble any particular volcano; rather it represents a class of objects. It has proportional similarity such that if it were enlarged

sufficiently, it would look like a prototype shield volcano, but it doesn't resemble any one mountain.

Like all models, scale models are necessarily simplified. The materials they are made from generally do not allow for much detail. As a result, they seldom function like their targets. A major reason they are functionally limited is that surface area and volume change at different rates when something is enlarged or made smaller. For example, if toy race car were enlarged to real size, it would be far too massive to race effectively, and if a real race car were simply reduced in size, it would be flimsy.

As a general rule, things that look similar but are of different sizes do not function alike. Certainly it is possible to make a scale model that looks like it is functioning similarly to its target; for example, motorized miniature airplanes may appear to function similarly to real airplanes, but the actual scale of the planes (for example, the thickness of the skin) is not the same, and the engines and control components are not strict miniature versions of the real things. Functional scale models generally must contain components that are not to scale in order to make them work.

Scale models portray only those features of the target that are needed for them to do their jobs. The addition of too many features makes the model more expensive and time-consuming to build without adding value for the user.

LIMITATIONS OF SCALE

A scale model has a simple concrete limitation. If an object changes in three-dimensional size, its mass changes by the cube, while its surface area changes by the square. In other words, if a model doubles in linear size in all directions (a factor of 2), its surfaces will increase by four times (2^2), while its mass will grow by a factor of 8 (2^3). For example, if we enlarge an adult 200-lb. human so he is double his linear size in all directions, he will weigh approximately 8 x 200 lbs, or 1600 lbs. However, the diameter of his leg bones and surfaces of joints, which are surface area functions, would grow only by a factor of four. Such an individual would not be able to function like the 200-lb man. The same principle applies to mechanical systems and models.

SCALE MODEL EXAMPLE

Take students outside to construct a scale model of the solar system, using balls of proper relative size to represent the planets and Sun. Where there is plenty of open space, station students holding the planets at appropriate distances from the Sun and each other. This is a great model for demonstrating the difficulties of distance and scale that astronomers face when studying the solar system. There are many functional relationships and characteristics missing from this model, however, most of which, if modeled, would make the model untenable.

Concrete Analogue Models

Concrete analogue models are models that represent functions in the target. They are less about the visual similarity of the model to the target, although this does not mean there is no similarity. In fact, model builders usually try to retain at least some superficial similarities with the target to facilitate recognition.

Most of the models listed in Table 3.1 are functional models; they demonstrate relationships in their target systems. Analogue models may be static or dynamic. Many experiments and activities students carry out in the laboratory involve the building of functional concrete models. For example, students may create a geyser by placing a funnel upside down in a pan of water and heating the water. The result functionally resembles a geyser but does not look like one, except in the most basic ways.

Another concrete analogue model is the mechanical model, standard in most Earth science classrooms, that shows the relative movements of the Earth, Moon, and Sun. The spheres representing the Earth, Moon, and Sun retain physical characteristics that allow us to distinguish them: the Sun is large and yellow; the Earth may be blue and smaller, and the Moon is usually smallest and white. These characteristics aid recognition, but they are not crucial elements of the model. What is important are the relative locations and motions of these bodies: They are intended to function—or act—like the target system. Clearly the system is not to scale and would be useless for teaching students what the system actually looks like.

Pictorial Models

Photographs, pictures, and drawings are concrete, two-dimensional models with the same properties as their three-dimensional counterparts. Photographs are intended to

exactly represent their targets and so are similar in function to scale models. Like scale models they are limited. Many systems cannot be photographed, because they are either too large (the solar system) or too small (the atom).

Diagrams and drawings may serve the same purpose as photographs, or they may serve the purpose of a concrete analogue model by representing relationships, usually at the expense of similarity in appearance. Drawings in a textbook of the solar system and cutaway views of the Earth usually have features that are not present in the actual target, but which are included to make the model useful.

Pictorial models differ from graphs and more abstract two-dimensional models, where lines and symbols represent relationships within the target. The idea of pictorials is to retain the similarity in appearance of the model and the target, so they can be interpreted generally without translation through a key set of symbols. In some models (three-dimensional wall maps, for example) features of the pictorial model may be combined with abstract features, so the lines defining various types of models are not always clear.

Pictorials, such as those found in textbooks, may confuse students in the same way as concrete analogue models do, because these models, and not the targets themselves, are what they will retain in their own mental models of a particular phenomenon. Some writers on the topic, for example, have blamed side-view drawings showing planets orbiting the Sun in extremely elliptical orbital pathways for perpetuating the misconception that summer occurs when the Earth is closer to the Sun and winter occurs when it is further away.

Distances may also be distorted in pictorials, in which, for example, the Earth and Moon may be perceived as much closer to each other than they are in relation to their sizes. Many of these misconceptions can be avoided by the simple expedient of analyzing the models—the pictorials—to identify the distortions they contain.

Concrete Models in Science Teaching

The job of scientists is to create models. It follows then that to teach science effectively we must engage students in creating models of many kinds. Since students are primarily concrete thinkers and learners, they are likely to benefit from creating concrete models. As a matter of fact, many of the lab activities we conduct in science classes call for students to build expressed models, including concrete analogue models and, to a lesser extent, scale models. When we include pictures and drawings as concrete models, it becomes clear that inquiry is a process of model building. This process parallels what scientists do.

Scientists often create concrete models in the laboratory to represent target systems in nature. They may create artificial earthquakes, tornadoes, volcanoes, waves, and similar phenomenon to test their predictions or to assess the behavior of the sys-

tem under different conditions. Wave tanks, for example, are used to test ship designs. Wind tunnels are used to test aircraft design and test the behaviors of buildings in high winds. Such models are especially valuable in cases where it is impractical or too expensive to work with the target itself.

Experiments in which a small, specialized system is used to represent a larger, more general target make use of modeling. Most of the experiments students do in the laboratory are of this type. An Earth scientist who studies the behavior of an accessible minor fault in order to generalize to a major but more inaccessible fault is using the minor fault as a model for the major one. Students do the same things in their lab experiences. Where we may be remiss as teachers is in not clearly defining the link(s) between the models created in the classroom and their targets, hence students do not fully comprehend the meaning—and limitations—of their work.

CONCRETE MODELS AND THE LABORATORY

Laboratory experiences are effectively about using and building concrete and functional models. An experiment has little meaning in itself; rather it must be related to some larger target to mean anything. For example, if students drop a cube of colored ice into a pan of warm water, they can observe the flow set up by the difference in densities of the cold and warm water. For this lab to have meaning, we have to relate it, as a model, to a larger target—such as the flow of ocean currents. The functional concrete model has similarities with and differences from this larger phenomenon that it is modeling. The importance of this lab experience is twofold: first, that students understand the ice pan activity as a limited model; and second, that science is a way of creating and using such models to build understanding.

Helping Students Build Their Own Functional Concrete Models

BEGIN WITH THE END

Stephen Covey, in his book *Seven Habits of Effective People,* advises readers to begin with the end in mind. This could not be truer than when engaging in scientific model building. An essential feature of such activity is the collection and evaluation of data. To build an effective scientific model, one must have an idea of what data are going to be collected and how they are going to be treated before beginning the construction. A construction firm would never begin work on a building without blueprints in hand. The builder must have the end in mind or risk wasting a great deal of time and resources. Yet we sometimes engage students in lab activities with no substantial end in mind, i.e., no real sense of the model we might expect to have built at the end of the exploration.

According to constructivist philosophy, students learn science by actually "doing" inquiry. When you guide students to construct their own models, they have a chance to participate in the full process of inquiry.

The science inquiry process is not simple for the novice learner. Science problems are not necessarily linearly structured: the problem-solving process is exploratory, iterative, and has no fixed method. Answering a question in science may use many paths and techniques, and the novice student may have no idea what they might be. Briefly, the challenges are that

- Learners do not know what paths and techniques are possible.
- Learners do not know how to perform the techniques.
- Learners do not know the rationales behind the techniques.
- Learners need support for organizing their work as they plan their investigations, track their work, and use their results to dynamically define the direction of their investigation. Many students have had little practice in logical organization.

Use the above list to decide how much scaffolding your students need as they design their models. Students may need help in sequencing their work, they may need examples or coaching, and they will certainly need feedback. Guided inquiry is a nec-

essary first step for all students. Initially, you, the teacher, will be your students' primary resource. As they become comfortable with the process of asking and answering their own questions—and presenting their work to others—your level of facilitation will change.

BUILDING FROM THE WORK OF OTHERS

Explanatory concrete and theoretical models are often stated as general principles that define the observed relationships. This inductive research is of the kind undertaken by early astronomers, geologists, and naturalists such as Darwin and Wallace. The models created by those who first explore a domain are essential to further scientific work, of course. The important point to make is that such explorations are purposeful, they are model-building activities, and the effective scientist circumscribes the models he or she builds. Darwin, for example, built a careful model describing the relationship among South American and Galapagos finches. He did not just explore wherever his attention happened to wander.

The Steps for Building a Model

Step One

The first step in conducting any research-based inquiry is to identify the guiding question we wish to answer. What is our goal? The question may not be well defined until later in the process, but it is important that the researcher have an idea of what it is he or she wants to know—a hypothesis.

Step Two

Observing and recording what we already know about our hypothesis gives us an initial sense about it—a mental model of it. But to explore our hypothesis more deeply we need to test it, often by building a physical model of the system being explored. In building, we try to relate the elements we know about to each other—stream flow and stream slope, for example.

Step Three

It's useful at this point to think of a title for your model. For example, to test the notion that igneous rock will be characterized by larger crystals when it cools slowly, state the purpose of the research as. "Creating a model relating crystal size to speed of cooling." Or, to look at ancient climates in your community: "A Model of the Occurrence of Fossils in the Layers of the Maple Avenue Road Cut." Finding a title first can actually help focus the model, and help us think clearly about the process we want to investigate.

Step Four

Once you know the nature of the model you wish to build, the next step is to analyze the task and figure out how to construct it. All meaningful research is preceded by a design phase, which may be linear or nonlinear. If my goal is to model the occurrence of fossils in a road cut, I may only need a systematic way to collect my data and record it. But I'll also need to plan how to collect and organize the fossils, and consider what I may want to look up in order to produce a better model (such as whether some layers contain different types and numbers of fossils than others). If my goal is to model the effect of slope on stream flow, I'll need to sketch out a plan and assemble a materials list.

Step Five

Next, identify the most important variables and separate them from those that are not important. Of those that are important to the model, it is essential to separate out those that must be measured, manipulated and controlled, or held constant, as the research demands. What variables do you plan to test, and why should they be included? What don't you plan to test? Record your thinking. Darwin only looked at a few distinctive features in his finches to construct his model, which he then was able to use as evidence of natural selection and evolution. If he had tried to catalog all of the variables in the birds' anatomies, his model would have been clumsy, huge, and have little explanatory value.

Step Six

Assemble your model. Practice working with your model until you can consistently collect data. Your testing helps you calibrate your model, and during this step you may make adjustments to the design or to the way you take measurements. Carefully document your work during this step. When you present your work to others for peer review, they will need to clearly understand the assumptions you've made, the variables you've chosen, and the uses and limits you've set for your model.

Step Seven

Next, run the model and collect data. Once we have data, we draw inferences. All of this is part of model building. Pieces are put together. Others are rejected. Gradually an integrated set of objects and relationships takes form, which constitutes the emerging model. Once you've tested and revised your model, put your work into a form to present for peer review.

PEER REVIEW IN MODELING

Scientists of all ages should know that unbiased testing of their proposed explanations by their peers can help make those explanations more robust, more possibly true. Peers test out explanations (or hypotheses or models) with their own data, to find out whether the results are within a realistic range. Professional modelers—those who develop the complex models for such agencies as EPA, NASA, and NOAA—depend upon a rigorous peer review system to help them adjust and refine their models of Earth and space systems.

Misconceptions and Concrete Models

The major problem with concrete models is that they become what students think of when we speak of the target. Students have never seen the solar system itself (nor have we), so the image they will evoke when asked about the solar system will be that of the model from which they have learned.

If we ask students to draw a picture of the Earth, almost invariably, the United States will be at the top of the picture and South America at the bottom. Why should this be? One rational explanation is that globes—in the United States, at least—represent the Western Hemisphere with North America at the top and South America at the bottom. There is absolutely nothing erroneous about placing a globe with the South Pole pointing up or pointing in any direction, for that matter. Directions do not exist in space.

The misconception that the Northern Hemisphere must be at the top of a globe is common and also politically as well as geographically misleading, because it implies a dominance of one half of the globe over the other. It is no accident that globes originated in the Northern Hemisphere and that this half is at the top. Additionally, "more important" places should be easier to find, and the "more important" places are in this half of the world. To some people, perhaps, the term "down under" implies that Australia somehow occupies an inferior place in the great scheme of things.

What might happen if we turned our classroom globe upside down, set it in a prominent place, and asked students to react to this new positioning? Perhaps some students would find it uncomfortable because the new position is unfamiliar, but there is also an aspect of thinking that it is wrong to have Europe and the United States spinning "upside down" at the bottom of the globe.

Most models don't have social-political implications like the placement of the globe does. Even so, models should be analyzed so students understand them to be representations of certain aspects—and only certain aspects—of their targets. The most important reason to use functional models is to explore relationships and behaviors. We should expect students to ask and answer the following questions whenever a functional model is used in instruction:

- What is the model for?
- What do the parts of the model represent?
- How is the model the same as its target?
- How is the model different from its target?
- How well does the model represent its target?

Often students do not know when they are building functional models. They do not make the conceptual leap to the targets. More likely than not, they will need direct intervention to gain these skills.

Analyzing Concrete Analogue Models

Most scale models are not going to create misconceptions, because students accept them for what they are. However, analogue models may cause misconceptions for the same reasons analogies do—because students do not understand which relationship between the model and target is being portrayed, and because students do not recognize the differences between model and target. To compare and contrast a model with its target, students need to directly experience the target, or else work with multiple models of the target. For example, students may conceptualize the structure of the Earth using a hard-boiled egg, a field trip to a road cut, photos of road cuts, core samples from a well digger, diagrams of Earth cross sections, hands-on sediment bottles, pictorial and written descriptions of earth's structure, and data of temperature at various depths of mines. Each model adds to students' mental model of the earth's structure, because their understanding of different dimensions of the target is being developed. As they study each model, students must be made aware that they are learning from models and should analyze what they have learned in that perspective.

Let's see how a teacher might guide students to analyze concrete analogue models.

John's class has been studying glaciers. John has them float two blocks of wood in water. He then has them pile sand on one block of wood. They observe and collect data on what happens to the block of wood. He also decides to ask questions about the model itself:

- What is the model for?

It shows how glaciers pressing down on the Earth's crust cause it to sink.

- What do parts of the model represent?

The sand is like the glacier—it's heavy. The blocks of wood are like the Earth's crust.

- How is the model the same as its target?

The block with the sand on it sinks lower in the water than the wood block without the sand, just like the crust of the Earth sinks under the weight of the glacier, compared to land without glaciers.

- How is the model different from its target?

The Earth's crust doesn't float in water, but lies on the mantle, which is molten material. The wood block isn't attached to anything around it, but land under a glacier is still attached to land around it. Is the magma displaced by the weight like the water is? Does it push up on the land around it?

- How well does the model represent its target?

Pretty well; it makes the point, but isn't very transparent. Would a better model be to stretch rubber over a box filled with syrup and then place a localized weight on it?

This analysis directly addresses the strengths and weaknesses of the model that the students have created. It may give rise to questions that otherwise might not occur to students: Does the effect of glaciers reach down to the mantle? What happens to the displaced material in the mantle?

Scientists don't build models just to explain things: Model building can itself lead to new questions, especially when a number of models are built to represent different aspects of the target.

Using Multiple Models to Teach Science

Since teaching and learning are processes of model building, the development of any conceptual model should include study of a variety of expressed models (as well as target systems, when they are available). Any one model usually shows only certain characteristics of its target. To fully develop a concept, students must use or build multiple models that portray various defining dimensions of the target. Concrete models are essential because most students are concrete learners, especially when introduced to new concepts. But we must also branch out to ensure that students are comfortable with models of many types, both concrete and abstract.

Let's suppose that we want our middle school students to develop a conceptual model of a tornado. First of all, we might elicit their ideas—their current mental model—of what a tornado is. We can then engage them in the study of tornadoes with the purpose of building a good conceptual model. A concrete analogue model appropriate for this task is the classic "tornado in a bottle," in which the whirlpool motion of the water in a bottle is related directly to the motion of the winds in a tornado. This model fits moderately well with an important—if restricted—characteristic of a tornado: that it is comprised of whirling winds.

The tornado in a bottle is a concrete analogue model. The whirlpool is a visual representation that is functional in an analogous sense, that is, the whirling waters carry debris in the bottle in a manner similar to—not the same as—the whirling winds of a tornado that carry debris from the land. Most of the dynamics of the fluid in the bottle are not the same as those of winds in a tornado. Driving forces are dissimilar, and in most other ways the model is not at all like the target. The tornado in a bottle is not a rich model, but it is a dynamic, hands-on one that can help students understand one very important quality of tornadoes.

Students may also study and dissect a cross-sectional diagrammatic model of tornado and discuss its parts (creating a verbal model). They may study and comment upon static pictorial models, read case studies (experiential models), read and analyze myths and legends related to tornadoes (literary models), and view a videotape of tornadoes (dynamic pictorial models) to gain a sense of the overwhelmingly destructive power of these winds. By studying a variety of models carefully selected to portray different aspects of tornadoes, students should be able to build an accurate and enduring conceptual model of this phenomenon. As they study each model, students should be aware that they are learning from models and should analyze what they have learned from each of them. Questions such as the following can help students in their analysis:

- What is the model for? What does it show?
- What are the parts of the model and what do they represent?
- How is the model the same as the target?
- How is it different? What is missing?
- Overall, how well does each model work? Does it add value to our desired conceptual model?

It is almost a given that students will construct a richer mental model of any phenomenon when they are required to interact with and/or build multiple models of it. Alternatively, if students build models sporadically and in isolation from each

other, they are not likely to gain the depth of understanding needed to internalize a given conceptual target. The richness of students' mental models will be directly related to the number and types of different concrete and abstract models of the phenomenon they are exploring.

Conclusion

Scientists regularly develop and use concrete models in research and field studies. Because conceptual learning is a process of model building, students studying science should be led to view their activities in class in this framework; that is, that they are actively building their knowledge rather than just acquiring it. Concrete models are essential to this process, since almost all students in elementary and middle school, as well as many in high school, are concrete learners. These students learn best if they can relate abstract models to those that are concrete. In fact, some researchers have found that all of us, including formal thinkers who deal well with abstract content, learn best concretely when we must learn something that is completely new to us.

It is important not just to engage students in model building—which many of us already do—but also to involve them in the conscious analysis of their models, whether concrete or abstract. That is, we as teachers must help students to become aware that they are building models, and that each of these models is limited in its power to describe a thing. It is only through fusion of multiple models that we develop sustained and useful knowledge of a targeted phenomenon.

Mathematical Models

Overview

Mathematical models deal with quantitative relationships. There are many kinds of mathematical models. Some are propositions, such as formulas and equations, while others are pictorial, such as figures, graphs, and pictograms. This chapter describes various mathematical models, with particular emphasis on their applications to science.

HISTORICAL IMPORTANCE OF MATHEMATICS IN SCIENCE

Mathematics fascinated the Babylonians and ancient Greeks, among others. To these early people, mathematics was not so much a way to solve practical problems, as a key to understanding the universe. Euclid's school, for example, imbued geometrical mathematics with mystical significance. Newton and his contemporaries felt that mathematics was literally a way of knowing the mind of God. In the mid-nineteenth century, when Darwin published his findings and conclusions on evolution, many scientists rejected them—not on the basis of the ideas themselves, but because the research did not include rigorous evidence based on mathematics. The trend toward qualitative research, which does not rely on numerical data, has been greatly impeded by continued insistence in some quarters that mathematical models are the only way to acquire and model scientific knowledge.

Historical Development of Mathematics

Every culture on Earth has developed some mathematics, and in some cases mathematics has spread from one culture to another. Although today mathematics is relatively similar across cultures, this has not always been the case. With roots in ancient Egypt and Babylonia, it developed rapidly in ancient Greece and was then translated into Arabic. At about the same time, mathematics of India was also translated into Arabic. Eventually, mathematics was translated into Latin and became accepted in Western Europe. Over a period of several hundred years, the Latin version became the dominant form of mathematics in the world community. Today the symbols of mathematics are known worldwide, and mathematics is an international language.

Historically, mathematical models have been incorporated into scientific work for hundreds of years. Early work in mathematics was quickly applied to solving practical problems and describing nature. While mathematics was certainly important in architecture and construction and in navigating, it was also an essential element in the mysticism of the classical period, especially for astrology. It is no coincidence that early work in science was conducted in physics and astronomy—two fields heavily reliant on mathematics.

Equations and Formulas

Mathematics today makes heavy use of formulas and equations, which are symbolic models. In an equation, two expressions are held to be equivalent—that is, of the same value. An equation sets up a logical relationship, as shown in the following examples:

$$2 + 2 = 4$$

$$(a + b) = 1$$

$$E_k = ½ mv^2$$

$$2H_2 + O_2 \rightarrow 2\,H_2O$$

Formulas are part of a larger and encompassing group of models that express relationships between variables. Equations are formulas, but the opposite is not true. Note the formulas given here, two of which are equations:

$$C_6H_{12}O_6$$

$$F = ma$$

$$\sqrt{(a^2 + b^2)}$$

$$(a + b)^2 = a^2 + 2ab + b^2$$

Formulas and equations are particular kinds of models—propositions—which are strings of symbols to which we give meaning. Consider the formula for force: $F = ma$, or "Force is equal to mass times acceleration."

The formula expresses a quantitative relationship relating force to mass, space, and time. We must interpret this model to use it, and to do that we have to understand and apply certain rules of correspondence. The symbol "=" has a precise meaning, as does the fact that the "*m*" and "*a*" in the formula are written side by side. By convention, we understand the operations we must carry out in order for the model to function.

The whole abstract model, $F = ma$, is not anything we can observe directly. It is an inferred expression of our mental model. We can see mass and view acceleration, but we cannot witness force (only its influence). Force exists only in mathematics. It is a good example of how humans literally construct reality.

Many of the concepts of quantum physics and astrophysics are mathematical, and they don't match observed entities in the external world. They are abstract conceptual models.

MATHEMATICS STANDARDS ON MATHEMATICAL MODELING

Standards published by the National Council of Teachers of Mathematics state: "In grades 9-12, the mathematics curriculum should include the refinement and extension of methods of mathematical problem solving so that all students can:

- Use, with increasing confidence, problem-solving approaches to investigate and understand mathematical content;
- Apply integrated mathematical problem-solving strategies to solve problems from within and outside mathematics;
- Recognize and formulate problems from situations within and outside mathematics;
- Apply the process of mathematical modeling to real-world problem situations."

Graphs and Pictorial Models

We use graphs, diagrams, and tables to show relationships between quantitative variables in an image format, rather than describing them as propositions. Graphs, pie charts, stem and leaf diagrams, and pictographs—among other devices—are ways to compare quantities and show change over time.

Graphs have elements that correspond to real-world variables and quantities. The line on a graph, for example, represents a series of related values in the external world. To read a graph, we must understand what each of its parts, individually and collectively, represents. The graph must be interpreted using relatively complex rules of correspondence.

Diagrams and pictorials, of course, have a long history in mathematics. They may be concrete or symbolic. A triangle drawn on a sheet of paper can be considered a concrete representation of a mental construct we call a triangle. It needs little interpretation, and has real-world correlates that look very much like it, such as a triangular sign. Graphs, on the other hand, require interpretation—there are no directly observable entities that look like the parts of a graph.

Models of this kind (graphs) are widely used in science. Students should be involved regularly and frequently in creating them. More importantly, students should be purposeful when they create them. Like all models, a mathematical construct is intended to do something: It serves a purpose and represents both what is in the student's mind and what he or she thinks is the external world. For this reason, it is as important to have students explain their models as it is to have them create them. It is important for us to determine whether students can explain what abstract models like graphs represent in concrete terms. Creating and explaining graphs require them to translate one kind of model into another.

Good sources of data in tabular and pictorial formats are available at several sites on the Internet. Of particular interest in Earth science are the NASA, NOAA, and U.S. Geological Survey sites. The model shown in Figure 4.1 is from a U.S. Geological Survey publication called Waterwatch. It can be found on the web at: *http://water.usgs.gov/waterwatch*.

Figure 4.1

U.S. map of real-time streamflow compared to historical streamflow for the day of the year (*http://water.usgs.gov/waterwatch*)

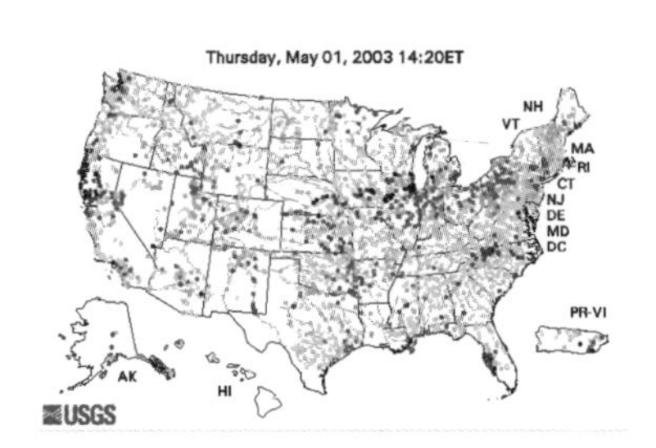

From this model, students may create a propositional model explaining how the data model was created and what it means in real terms to people in different regions. As an introduction to weather and climate patterns, this model could be one included as an element in a larger conceptual model of weather patterns in the United States that we would like our students to construct.

Statistical Models

It is difficult to practice or interpret science today without at least some background and understanding of statistical model building. Statistical models are abstract constructions concerned with the treatment of data. We use these models to make inferences, confirm predictions, and make decisions on problems. Usually, statistical models involve samples that are smaller than the whole target about which inferences are to be made. Many problems must be resolved to make sure generalizations are valid.

Whenever possible, students should have opportunities to create models based on data they have collected. To that end, we should

- Select or develop activities whenever possible that require students to collect and evaluate meaningful data.
- Pool class data for analysis, rather than asking individuals to process only their own data.
- Encourage students to discuss the data and search for patterns. Emphasize that the patterns should lead to conclusions rather than beginning with conclusions and forcing the data to fit them.
- Make sure students understand the basic assumptions and processes of analyzing data.

Most of the science activities we do with students do not need the kind of complex data analysis scientists use. This does not mean, however, that we cannot help our students understand some of the major ideas underlying the interpretation of their data. Some important points to make and reiterate are the following:

- There will always be variability in the data that are collected.
- The data are only as good as the design of the research and the care taken to collect them.
- All other things being equal, larger samples give us more confidence in our results than smaller ones do.
- A difference in two means does not automatically indicate a difference in the groups being tested. Other factors to consider are the sample size, the precision of the

research and observations, and the amount and clustering of data collected.

- A data point represents conditions at the point in time when it was recorded. If environmental factors change, the old data may no longer represent the current system.
- Whether qualitative or quantitative, the basis for good science is good data. The model created from the data should be as clean and clear as possible.

Since the purpose of science (and statistics) is to learn about a target and not just about a sample, students should be required to discuss and evaluate the fit of their statistical models to larger targets.

The validity of a generalization, of course, depends upon the assumptions we make about our sample. Suppose we take our students outside to study soils. They dig equivalent holes in three different locations in clay soil and loam soil, then collect data on how quickly a liter of water poured into each hole soaks into the soil. To what extent can we generalize these findings on permeability to all clay and loam soils? What variables in other locations might alter these findings?

Applied Mathematical Models in Science

When we apply mathematics to solving problems in science, the models we create are subject to the same rules of evidence as apply to any other scientific models. The schematic shown in Figure 4.2 shows the process of using mathematical models to solve real-world problems.

Figure 4.2

Using mathematical models to solve real-world problems

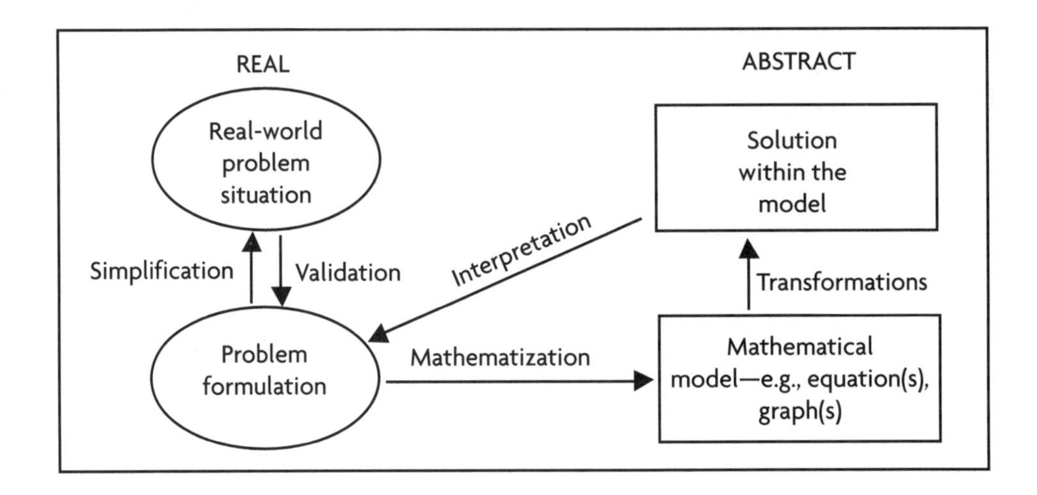

Most of our models of objects and relationships in the fields of quantum physics and theoretical astronomy originated from mathematical logic rather than scientific evidence. We can't see or experience phenomena in these domains; rather we must rely upon predictive models derived from them. We can then accept the validity of these mathematical models when we show we can consistently predict from them.

For example, string theory proposes that the smallest entities in the universe are things called strings. Strings vibrate with different frequencies, which accounts for different behaviors and charges in subatomic particles and the forces that unify them. The term "strings" is an analogous invention. Strings have no dimensions and are unlike anything we can see in the macroscopic world. They are mathematical constructs, not real entities in any way we can understand.

Models such as strings are hypothetical until they can be shown to have predictive power. For example, when Einstein predicted from mathematics that gravity should bend light, his idea remained a hypothesis until a way was devised to demonstrate that a massive object could bend light.

Proposed models grow into theoretical ones (we speak today of string theory) when they gain enough explanatory power in relation to accepted models to become accepted themselves. Even so, if they are eventually found to be inconsistent with observation, they will be modified or dropped, as the occasion demands.

Building an Applied Mathematical Model: An Example

Naturalists in parks often make statements about the projected population of some species of animal a few years down the road. It's not unusual to wonder, "How do they know that?" This activity, taken from the NCTM web site, models some of the math used to make these projections.

Biologists compile data on the death rate of animals and can make predictions about the size of a cohort of animals over a period of time. (A cohort is a group of animals all born at about the same time.) Students can build an applied mathematical model showing the survivorship of a cohort of animals. Here's how:

Organize students into groups, and give each group a sturdy bag containing 100 pennies or 100 two-color counters. The pennies (or counters) represent a cohort population of 100 animals. With these materials, we can assume that each animal has a 50 percent chance of surviving the year. Next, have students shake the bag and drop the pennies (quietly!) on a desk or table. Any penny showing tails represents an animal that has died and should be removed. Ask students to record on a chart the number of remaining pennies, then put them back into the bag. Students should repeat this process five times (representing five years in the life of the cohort's existence). Finally, have students graph the data in their charts and discuss what they notice about the population changes over time. Older students can use a graphing calculator to fit a

function to the data. Do students think the population will ever reach 0? If a large quantity of number cubes (dice) are available, have students repeat the experiment using 50 or 100 number cubes and removing the 1's as animals that have died each time (an 83 percent survivorship rate). Students can discuss the possible limitations of this mathematical model.

A THEORY BASED ON MATHEMATICAL CALCULATIONS

The concept of black holes was predicted by mathematical calculations—with no empirical evidence for their existence. For years, the scientific community regarded the notion of black holes with skepticism. Only with the advent of orbiting telescopes have astronomers witnessed phenomena consistent with the existence of black holes. Observations of the behavior of matter and emissions of high-energy radiation from areas of space thought to harbor black holes are consistent with predicted behaviors that might be expected from such a body. The notion of black holes has gained theoretical standing over time.

Pure Mathematical Models

The term "pure mathematics" refers to the creation and manipulation of mathematical models through internal rules of logic. We can contrast this idea with applied mathematics, in which models must be validated by empirical observation. The purpose of pure mathematics is to discover quantitative relationships, using rules that the community of mathematicians understands and accepts. In pure mathematics, written formulas represent mental models—nothing more. They are internal ideas represented as symbols.

Although mathematics underlies much of science, purely theoretical mathematical models are built on a foundation different from scientific models. In pure mathematics, mathematicians create logical mental models that do not require empirical proof for acceptance. On the other hand, mathematical models used in science must be shown to be empirically valid.

WORKING WITH PURE MATHEMATICAL MODELS

Project Interactivate—funded by the National Science Foundation and the Department of Defense Education Activity—is a freely available set of java-based courseware for middle school mathematics explorations. "Interactivated" lessons, discussions, and activities provide opportunities for students to visualize and experiment with pure mathematical models of algebra. *www.shodor.org/master/interactivate*

Using a Pure Mathematics Model to Solve a Problem: An Example

The following problem was first posed in the ninth century and finally solved 800 years later in 1654 by the famous French mathematicians Pierre Fermat and Blaise Pascal. Note that in this version, we have modified the content but not the nature of the original problem setting.

Real-world problem situation.

In a two-player game, one player wins a point at each toss of a fair coin. The player who first attains n points wins a pizza. Players A and B commence play; however, the game is interrupted at a point at which A and B have unequal scores. How should the pizza be divided fairly? (The intuitive division, that A should receive an amount in proportion to A's score divided by the sum of A's score and B's score, has been determined to be inequitable.)

Problem formulation.

Consider the situation with the following data: The winning score is $n = 10$. When the interruption occurs, the score is A:8 and B:7. The pizza will be divided in proportion to each player's probability of winning the game.

Mathematical model.

At each turn, P(A wins a point) = P(B wins a point) = 1/2. A's share = P(A wins 10 points) x area of pizza; B's share = total pizza – A's share. Let a square region represent the original game state with the score A:8 to B:7 as indicated. At each turn, the square

or interior rectangles are halved to represent P = 1/2 for winning (or losing) a point. Thus, in this model, the resulting fraction of the original area also represents the probability of reaching that game state.

Figure 4.3

A mathematical model showing how the pizza should be divided

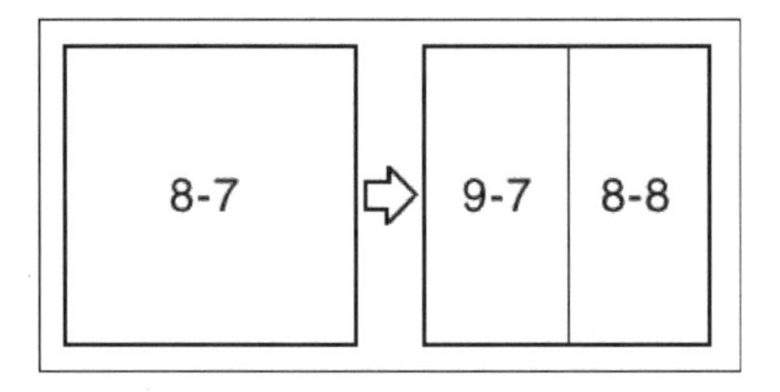

Interpretation of solution in original problem formulation.

A's share = 11/16 of pizza
B's share = 5/16 of pizza

Teaching with Math Models: An Example

The following sequence models the way we build and use math models to answer science questions. We have adapted it from a project developed by NASA and NCTM to illustrate the Mathematics Education Standards. The project can be found at *http://illuminations.nctm.org/lessonplans/9-12/deris/index.html.*

Posing the Question: Space Debris—Is It Really That Bad?

The accumulation of space debris in orbit around the Earth poses a problem for NASA and space scientists from other countries. Such debris includes payloads that are no longer operating; spent stages of rockets, assorted parts, and lost tools; debris from the breakup of larger objects or from collisions between objects; and countless small pieces, such as flakes of paint and even smaller items. Because bodies in Earth's orbit travel at approximately 17,500 miles per hour, a collision with even a tiny object can have catastrophic effects.

In 1990, scientists estimated that a total of 4 million pounds of debris was in Earth's orbit. At that time, they also estimated that we were adding 1.8 million pounds per year, which in a few years would result in 9.5 million pounds of orbital debris. The 1990 prediction also anticipated that the amount of debris being added would increase to a rate of 2.7 million pounds per year by the year 2000.

Activity 1

1. How much is 4 million pounds of anything? Or 9.5 million?

2. Give at least three concrete examples that would help another person get a sense of how much 9.5 million pounds of debris is. For example, finish the following sentence: A total of 9.5 million pounds of pennies would fill _____________.

Modeling the Problem: Linear Growth

We cannot solve this problem directly, because we cannot locate, count, and weigh all objects in orbit. Nor can we predict with assurance when any two objects will collide. Instead, we must rely on mathematical models to help us represent the problem and identify trends and expected outcomes. In these activities, students create and compare various mathematical models to help them investigate some of the questions raised by the proliferation of orbital debris. These models are greatly simplified in their assumptions so that students can investigate them with calculators, spreadsheets, and graphing utilities. Nonetheless, the models provide insight into the process of mathematical modeling and its importance.

Activity 2

1. Use the reported 1990 rate of increase of 1.8 million pounds per year and write a linear model to predict the number of pounds of orbital debris at the end of any given year, t. Assume that $t = 1$ represents 1990.

2. Write a second linear model using the predicted 2.7 million pounds per year rate of increase and the initial 4 million pounds for 1990.

3. Evaluate each model for several years to determine the year in which the predicted 9.5 million pounds of accumulated debris would occur:

   With the first model ____________________

   With the second model ____________________

4. When a vehicle travels at a constant rate, r, for a length of time, t, the distance, d, is modeled by a linear function. Compare the familiar linear model for distance-rate-time with your linear models for accumulating space debris. Why can we refer to your linear models as "constant-velocity models for amassing space debris"?

5. Do either of the linear models accurately represent the situation of escalating

amounts of space debris described in the section "Posing the Question"? Why or why not?

Refining the Model: Quadratic Growth

Does either rate, 1.8 million pounds per year or 2.7 million pounds per year, describe how much debris was building up between 1990 and 2000? Which rate of increase should we use? Obviously the amount being added each year changed during that period, but by how much each year? The problem is one of acceleration, not constant velocity, so we need to adjust our model.

Again, let's make the simplest assumption: the rate at which debris is added increases at a constant rate from 1.8 million pounds per year in 1990 to 2.7 million pounds per year in 2000. This change means that over the ten-year period from the end of 1990 through 2000, the rate (velocity) of littering will increase by 0.9 million pounds per year ($2.7 - 1.8 = 0.9$). We will also assume that this increase is achieved in equal annual increments of 0.9 million pounds per year in each year of the decade.

Activity 3

1. Complete a table to show the amount of debris added each year and the total amount in orbit at the end of the year.

2. Assume that the increase in the velocity of littering is achieved in equal annual increments. Write a linear equation that describes the increase in the amount of debris being added each year—that is, the increase in the annual velocity of littering as a function of the number of years since 1990. In this case let $a = 0$ in 1990 based on the assumption that the 1990 rate of 1.8 million pounds per year is the baseline rate. Then $d = f(a)$ represents the rate of littering a years after 1990.

3. The situation described in the equation, where the rate of increase of litter is itself increasing at a constant rate, is analogous to a vehicle that accelerates at a constant rate from an initial velocity, v_0, to a final velocity, v_f. Use the data from the table you developed (Step 1) to create a scatter plot of the total number of pounds of orbital debris that have accumulated relative to the year. The graph should cover the period from 1990 through 2000.

4. Fit a line to the data and decide whether the accumulation of debris appears to be linear. Write a conclusion and describe the evidence on which the conclusion is based.

5. Use a graphing calculator or computer graphing program to calculate the linear-regression equation for these data. Do the calculations support a linear relation-

ship? Explain. How does this line compare with the line that was fitted manually?

6. Generate a quadratic-regression equation for the same data. Write the quadratic-regression model. How well does this equation fit the data, compared with the linear approximation?

7. Compare the quadratic-regression equation with the two linear-regression equations developed earlier.

8. In each case, use the models to predict the accumulation of debris after 20 years, 30 years, and 50 years. Describe the behavior of the linear model versus the quadratic model over time.

9. For the period from 1990 to 2000, the graph of the quadratic model lies between the graphs of the two linear models. Explain why this result is reasonable. Will the quadratic graph always lie between the two linear graphs? Explain.

10. Explain why the quadratic model for the debris problem can be described as a "uniform acceleration" model.

Extending the Models: What Goes Up Might Come Down

All the models developed thus far assume that every year some of the debris slows down enough to re-enter the atmosphere where it burns up or, on rare occasions, returns to Earth. Assume for the moment that 10 percent of the debris in orbit at the beginning of any year will be destroyed during that year.

Activity 4

1. Modify the linear, quadratic, and exponential models to account for the situation in which additional debris is being added each year while 10 percent of what was already in orbit is being destroyed.

2. In which case—linear or quadratic—does the assumption of a 10 percent re-entry rate have the greatest effect?

3. Assume the same rates of adding debris as previously, but try different rates of annual destruction of orbital debris. In each situation, does a destruction rate exist that will result in a net decrease in orbital debris, despite the fact that additional debris is being added? What might be some advantages of knowing if such a rate is possible?

Putting the Models to Work

Mathematical models are powerful because they lead us to ask "What if?" questions. An earlier question is an example: What if we could increase the rate at which orbital debris is destroyed? Other questions might include these: What if we decrease the rate at which we are adding debris and find a way to increase the rate at which existing debris is destroyed? Such questions lead to open-ended investigations using mathematical models.

Conclusion

In this chapter, we read that mathematics and science, though related, are not the same. Mathematics is one kind of model people use to explain and understand their world, but it is by no means the only one, or even the best one for many purposes.

We can construct scientific models in some cases without mathematics, but mathematics is nonetheless an important tool for scientists. As we have developed our ideas about the probabilistic nature of reality, and as our measurements have become more refined, we can understand why the science community has become increasingly reliant on statistics for decision-making. If our students are to become scientifically literate citizens, they must have enough knowledge of statistics to at least ask the right questions when making statistics-based decisions. As teachers, we can help students grasp the basic ideas in statistics and use them regularly in their science work, beginning with simple ideas and progressing in grades to more complex understandings.

Students should understand that the data scientists collect seldom lead to models that are completely clear. Errors of measurement and differences caused by uncontrolled and uncontrollable variables require that scientists make choices about what and what not to include in their models. The same caveats apply to students' data.

Students must also understand that a model is a representation of a system—imperfect by definition. However, the tool of statistics can guide scientists (of all ages) to collect an amount and type of data that will closely approximate a system. The tool of computers can allow precise and iterative testing and adjustment of the algorithms within a model. Peer review can provide the modeler with new ideas to improve a model. So students should recognize that modeling itself is a learning process.

CHAPTER 5

Computer Technology and Modeling

Overview

In the days before computers, scientists built outcome models variable by variable, using thought experiments and the data they were able to collect from methodical research. They engaged in thought experiments to anticipate outcomes and identify new directions for their research, but were unable to grasp many complex systems because of the large number of variables involved. The analytical tools of scientists were not up to the task of modeling and processing many different variables at once.

With the arrival of computers, scientists became increasingly able to deal with complex systems. Not only could they test many more combinations of variables than they had in the past, but they could also manipulate variables in their models to create predictions that they could compare to observations in the field, gaining insight into phenomena that had previously defied explanation.

For example, the tremendous advances in weather forecasting in recent years have been due largely to computers capable of collecting and processing huge amounts of data very rapidly. This feature has allowed meteorologists to achieve insights into global and atmospheric behaviors that were not possible several decades ago.

Supercomputers today use parallel processing to enable scientists to work on problems of enormous magnitude. In general, a machine using a given number of parallel processors can solve a problem that number of times more quickly: A machine with five parallel processors will work five times more rapidly than one with a single pro-

cessor. As of this writing, the largest modern supercomputers have over 9000 processors in parallel. Software now available allows volunteers linked to the Internet to leave their machines on for access by research institutions at night. Researchers at these institutions can then link many machines together as parallel processors, so they can work on complex problems while the volunteers' computers would otherwise not be used.

Computers and satellites together facilitate the collection and processing of data from many points on the globe. Anyone can find excellent web sites discussing the use of computers in weather forecasting by typing in a search using "computers" and "weather forecasting" as key words. A good site currently available is: *www.usatoday.com/weather/wmodels.htm*. This site introduces forecasting with computers and describes several models used in predicting the weather. Other information is available at sites sponsored by the U.S. Geological Survey, NASA, and the National Weather Service.

Computer models, as helpful as they are, still must meet certain requirements to be accepted as scientific. That is, they must be confirmed by observation and prediction of empirical events. Computer models also have the limitations of any models, so they should not automatically be accepted as valid. In fact, they are only as good as the information and assumptions that they are given.

Building a Computer-based Model

A computer-based model uses the conceptual framework of computational science—application, algorithm, and architecture:

- The *application* is a concise statement of the problem to be solved.
- That problem must then be translated into a mathematical format, known as the *algorithm*.
- The algorithm must then be converted into an appropriate *computer code* using a programming language to be run on some piece of hardware.

Scientists must begin with a concept of the system they wish to model; that is, they must have a certain set of assumptions upon which their model will be built. If these assumptions are erroneous, the entire model may also be in error, even if it is logically consistent. Scientists must also identify the variables relevant to their model and the relationships among these variables that may alter the outcomes of their interactions. The next task is to express these relationships mathematically, as algorithms. Each variable acquires a beginning value, which may be a set quantity or a formula relating the value to values of one or more other variables. The model is then translated into a

computer language and run, producing some sort of output such as a graph, table, or pictorial representation.

One thing that becomes clear to anyone running dynamic models is that the models can become quite complex. Models that seem simple in static drawings are much more difficult to work with when they are interactive. In textbooks, for example, the water cycle model looks like a simple system, but when we begin to construct a dynamic model of it, many values and variables emerge. If nothing else, model building with computers can help us and our students appreciate the complexity of natural systems.

THE LIMITS TO GROWTH

In their 1974 book, *The Limits to Growth: A Report for the Club of Rome's Project on the Predicament of Mankind,* Donella Meadows and others designed a series of computer models describing major shortfalls of important fuels and minerals before the turn of the twenty-first century. The book warned readers that unless changes were made in policy and growth very soon, modern civilizations were doomed. Many of the projected deadlines have come and gone, and in most cases, the projections have turned out to be inaccurate. While the basic assumptions behind the models were logical, many variables were not added into the model. This is not to say that the difficulties projected by the report will not someday occur unless changes in the use of resources are made. It is merely to point out the potential inaccuracies in computer modeling of future events when complex systems are involved.

Computer Models for Teaching

Computer models for teaching are becoming increasingly common. They are often used to model events that

- occur too quickly or are difficult to observe (such as formation of a raindrop);
- occur too slowly (such as tectonic movements in the Earth's crust);
- are too costly to replicate in the laboratory (wind tunnel modeling); or
- are dangerous (rapid combustion experiments).

With computer models, students can simulate events, modify experimental vari-

ables, and repeat events under different circumstances to gauge the effects of such changes. This hands-on learning gives students rapid feedback on their experiments and helps to develop scientific intuition.

Whenever possible, students should support their computer models with models created in other media, especially hands-on concrete models and data collection similar to the suppositions used to make the computer model. This follows from the multiple models ideas expressed in Chapter 3. As teachers, we must remember that computer modeling is only one of many ways to teach processes and concepts. Students still need to spend time theorizing, researching, and reading background material to support the guiding question of the computer model.

In any modeling process, we all need to keep in mind the old computer acronym, GIGO (garbage in garbage out). Models are only as good as the data and assumptions used to create them. Like other models, computer models must be carefully analyzed for their "goodness of fit" to the system they are representing.

The greatest challenge in computer modeling is choosing the appropriate tools for a particular problem. Students should understand the strengths and weaknesses of various computational tools, and the impact these tools have on answering the key question that all computational scientists ask: "How do we know the model is right?"

POSITION PAPER ON CLASSROOM TECHNOLOGY

According to the Council of State Science Supervisors, "The use of technology in science education is more that just having a computer and accessing the Internet. Science education relies on the use of scientific technology. Students can best understand science by using computers, the Internet, calcularo based laboratories (CBL's), temperature probes, projection panels, telescopes, microscopes, pH meters, Geiger counters, rheostats, oscilloscopes, lasers, air tables, spectrophotometers, glassware, VCR's, graphing calculators, electronic balances, and the other tools of science in observing, classifying, inferring, and experimenting in class. A science class without the tools of science cannot adequately practice science."

The Council goes on to state that *ideally,* each science teacher needs a computer, projection panel, overhead, and scientific probes to demonstrate science programs or experiments. Five additional computers are recom-

mended as a minimum for student work for a classroom to begin to compete with the better 7–12 schools.

Regardless of its type or complexity, technology is a tool. Its effectiveness as an instrument of scientific learning is derived from the teachers and students who use it. The criteria for the success of technology in the science classroom relates to the following questions: Is it to improve student learning? Is it to increase the effectiveness and efficiency of science teachers and to expose students to the technology that is or will be the norm in their work or private lives? Each state department and local school district needs to develop a plan of how, when, and what to purchase in scientific technology over time.

Using Computer Technology to Teach Science

Using the computer to create models does not always involve creating computer models per se. There are many software programs—from word processing packages to complex analytical tools—that students can use for various reasons. We should provide students with experiences in the use of this technology for

- presenting their constructed models to others;
- finding data and resources to use in their models;
- interacting with others to share and exchange data;
- collecting and processing real-time data in the lab and field using probes;
- modeling phenomena with or without user input in a tutorial format;
- modeling phenomena in inquiry format by manipulating set variables;
- constructing and testing open-ended models using pre-defined objects; and
- constructing models from real data on the Internet.

Presenting Models to Others

Influential scientists must be able to present convincing evidence and arguments to others. These skills are usually developed in a variety of school subjects and should be practiced as well in science—and here technology can play an important role. The

following applications are familiar to most of us and we will merely identify them. Our students should be aware of these resources and we should require that they use them as often as possible.

- Several commonly available software programs permit students to make slides that can then be either projected or printed to plastic transparencies.
- Word processing programs allow students to print out user-friendly, easy-to-read graphs and tables using templates.
- Photographs can be transferred to computers and presented or printed directly from them.
- Computers allow students to write and edit papers and presentations more easily than was possible in the past.

Unfortunately, the Internet has also made it easy for students to download someone else's work and present it as their own. To prevent such plagiarism, assignments should address unique questions and teachers should either insist on notes (as well as a final product) or monitor the development of the model as it is created through a series of steps.

Data Collecting and Processing

Just as calculators have the potential to enhance the learning of mathematics, probes for data collection and processing software have the potential to enhance the learning of science through the creation of computer-based data models. A number of educational suppliers have probes available for measuring and recording variables such as temperature, pressure, motion, and electromagnetic energies. Probeware is generally usable with different computer platforms, and comes with relatively inexpensive software. This software allows students to examine changes in the variables over time as well as to store and print graphs and tables of results.

However, as calculators in mathematics are not intended to supplant a basic understanding of how to construct mathematical models, programs for collecting and processing data are not intended to replace activities designed to ensure understanding of computer models. When we make allowances for questioning and discussion to ensure understanding, it then becomes apparent that we can encourage the use of technology to facilitate inquiry and model building.

Probes have the advantage of responding more rapidly to changes than most traditional instruments, and the readouts are easier to observe. "Real-time" graphing on the monitor enables students to see the results of their manipulations as they occur, adding both interest and speed to the experiment. There is some evidence that the

proximity of manipulation and results helps students to see and better understand the causal connections in their models.

As an example of their use, consider an experiment in which we insert three temperature probes at intervals into a heavy cardboard cylinder, such as a mailing tube. Once we have the probes attached to a computer, we can take a flask of cold air from the refrigerator and pour it into the upended cylinder. As the heavy cold air settles into the cylinder, data can be collected to show that cold air is heavier (more dense) than surrounding air. This is much easier and more efficient than reading thermometers, and we have a record of the changes on the computer.

Contemporary probeware is available for use with PDA handheld devices and calculators, as well as with computers. This adaptability makes probeware practical for field trips, where data can be stored for later retrieval and analysis.

In addition to probeware, instruments such as digital cameras are increasingly affordable and allow images to be downloaded directly to image processing software. For qualitative research studies, these kinds of data may be extremely valuable. In addition, images are very important for communicating science findings to others.

Tutorials

Tutorial software ranges in type from passive to interactive presentations that engage students in construction of a model from stock elements. The purpose of tutorials is to teach students directly, rather than engage them in exploration.

Passive presentations may not engage students in anything more than clicking a button to continue the program. They are the equivalent of reading a text. While passive tutorials have their uses and may be entertaining, they have some of the same shortcomings that a passive lecture has—that is, they may fail to engage learners in constructing their own conceptual model of the phenomenon being considered.

Active tutorials may involve students in the model-building process—sometimes through the development of visual analogue models. An active tutorial in chemistry, for example, might engage students in the drag-and-drop construction of molecules using icons to symbolize electron pairs and elements. It may then allow the electron dot diagram to be transformed into a three-dimensional representation that can be rotated for study.

Tutorials usually have only one correct outcome and thus do not facilitate inquiry. However, they have the advantage of actively engaging learners in construction of their own conceptual models.

Computer Modeling in an Inquiry Format

Inquiry occurs when learners construct their own models by observing a phenomenon, collecting data, and making sense of that data, either independently (open) or with the assistance of the teacher (guided). Software that facilitates inquiry is sometimes used in place of certain expensive or otherwise difficult hands-on laboratory experiments, and for acquiring data that are impractical for students to gather on their own.

Students should use models of this kind both to learn the content and to develop understanding of the processes of computer modeling. The simplest of these software programs require only that students enter values for variables. The algorithms are fixed. The variables are already linked to each another, and by changing values of variables by turn, students can observe changes in other variables. For example, a model of this type might demonstrate that an increase in sunlight falling on a lake increases its evaporation rate.

Last Straw Project

An example of computer modeling in an inquiry format is the Last Straw Project, funded by the National Science Foundation. It is located on the Web at *http://cycas.cornell.edu/ebp/projects/laststraw/ise.info.html.*

This pilot project is a combination simulation and tutorial on climate and plant water-use characteristics. It is an interactive simulation model that allows users to experiment with how plants deal with drought. Users also have access to a self-paced computer tutorial that covers basic plant structure, water use, gas exchange, and adaptations to extreme environmental conditions. The model allows students to compare the growth of plants in two different chambers and to select different variables for each chamber. A series of guided questions aids the user in the analysis of the results.

MASTER Tools

The Shodor Education Foundation web site (*www.shodor.org*) makes available a number of computer modeling activities suitable for use in the Earth science classrooms. The MASTER Tools—developed by the foundation, in collaboration with the National Center for Supercomputing Applications, George Mason University, and other education organizations—are interactive tools and simulation environments that involve students in exploration and discovery through observation, conjecture, and modeling.

According to the Shodor Foundation, the goal of the program developers is to integrate MASTER tools with new collaboration tools and online research facilities to create an authentic scientific experience. The simulations and supporting curriculum materials are, and will be, designed to comply with the National Science Education Standards and the National Math Education Standards.

Instructional materials currently available are GalaxSee, SimSurface, and the Fractal Microscope. Each program comes with detailed lesson plans, objectives, background information, and its relation to the National Science Education Standards. In the GalaxSee program, for example, students simulate the creation of a galaxy. In doing so, they work through issues of computer modeling and respond to the kinds of decisions scientists make. The company is currently developing materials in environmental science, as well as in other areas of science.

Student-Mediated Construction of Models

Another class of computer models involves students in actual construction of models from icons representing variables. These sophisticated programs do not limit students to preprogrammed algorithms but allow students to construct their own. Two widely used programs are Model-It and STELLA.

Model-It

Model-It—available from the Center for Highly Interactive Computing in Education (*www.hi-ce.org/index.html*) at the University of Michigan—is an open-ended program, perhaps most suitable for middle-level and secondary students. The program provides for the identification of objects, variables, and relationships within a system. For example, to construct a simple system to test variables related to the water level in a pond, students can create objects labeled "volume of the pond," "precipitation," "groundwater," and "evaporation." Note that these objects are similar to the nodes that make up a concept map.

In this program, students select objects, then assign each a numerical or qualitative value. For example, students may choose a number to represent the amount of rainfall or water in the pond. Alternatively, students may simply elect to use relative qualitative terms such as "high," "medium," and "low." Once the values are set, students build the relationships between elements of the model by linking the objects with arrows and specifying the relationship on a pop-up window.

Students may then run the model with their selected values and observe the results on a graph as it is plotted over time. In addition, students may add and observe a meter window marked "dependent" or "independent" for each variable. The independent variables may be manipulated even as the system is running, allowing students to observe changes in the graphed relationship(s). The pop-up windows have spaces in which students can justify their beliefs about the relationships among the variables.

Figure 5.1

Example of running simulations on Model-It
(http://goknow.com/Products/Model-It.html)

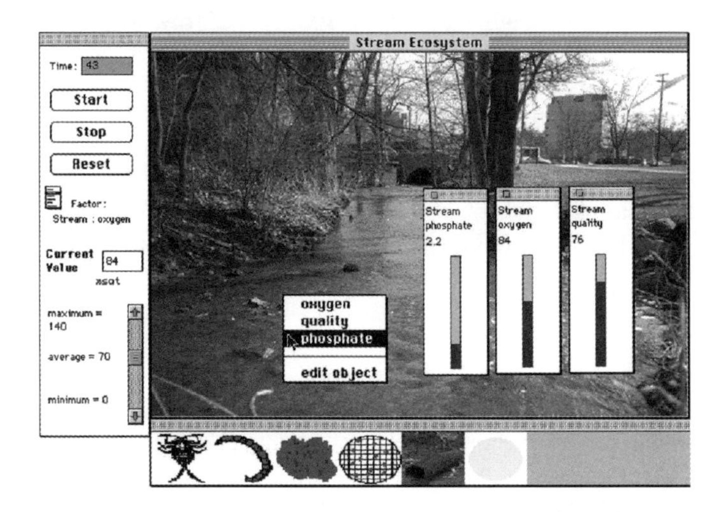

Model-It comes with icons with pictorials that can be used as objects for systems to study soil, water, communicable disease, air, and decomposition. It also includes a generic icon that can represent objects not available in the program package. In addition, users can import images and include them in the model. Because any number of icons or pictures can be used to create objects, Model-It is a very user-friendly package.

As students select objects, variables, and relationships, the program uses these to construct the mathematical relationships underlying the model, so students need not grapple with complex equations. On the other hand, Model-It is limited in its functional range. For example, it does not allow for constants in the model. Relationships must be increasing or decreasing. Hence, the constant loss of water from a pond due to human use, for example, cannot be part of a model. These kinds of restrictions are a problem with models that do not allow users to access the underlying mathematics.

Figure 5.2

Example of defining absolute relationships with Model-It
(http://goknow.com/Products/Model-It.html)

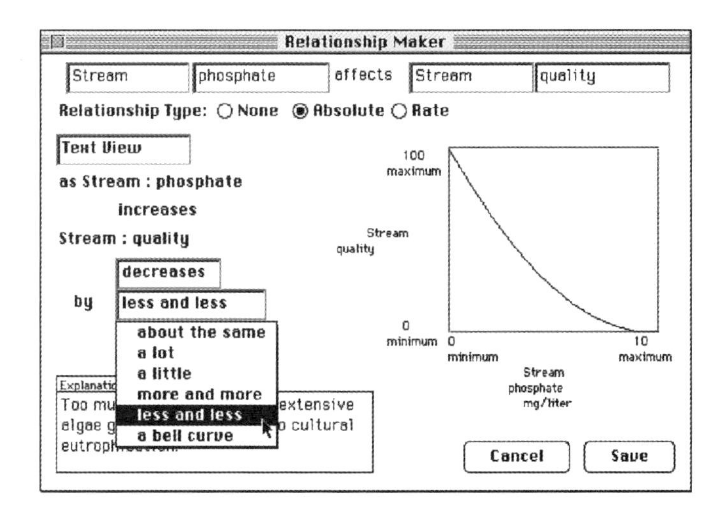

STELLA

STELLA is commercial software marketed by High Performance Systems, Inc. It allows users to construct models from five elements: stock (things that accumulate); flow (input and output to the stock); converters (factors that influence flow and stock); connectors that knit these elements together; and decision diamonds. An overview of the program is located at *http://archive.ncsa.uiuc.edu/edu/RSE/RSEindigo/stella.html.*

STELLA allows for the development of complex models. It is the basis for a number of relatively simple data-entry models that allow learners to manipulate values of variables but not the components of the system being modeled. Unlike Model-It, STELLA is not strictly an academic program. Beyond educational uses, it has the power to model applications in science, business, and industry.

STELLA is more difficult to learn than such programs as Model-It as it requires students to create mathematical expressions as they add converters. While this dramatically increases the range of possible models, it limits model building to students with enough mathematical understanding to complete the algebraic equations needed to calculate complex variables. By and large, this program has been used successfully with high school students; it allows for much more flexibility in the modeling process than that of easier-to-use software packages like Model-It. A trade-off between ease and flexibility in model building still exists.

On the other hand, in middle schools, STELLA may prove an effective tool for guided inquiries with small groups, in which the teacher facilitates the development of the model with student input. The windows used by STELLA make selecting the variables influencing values less complex than they might otherwise be by identifying the variables that must be included in the equations.

SESOIL

SESOIL (an acronym for the SEasonal SOIL model) is an online EPA model. It can be accessed at *http://igems.shodor.org/igems/*. The model simulates movement (transport) of a chemical in unsaturated soil. It models the effects of user-described characteristics for variables associated with climate, the soil column, and the chemical of interest. It models the effects of environmental fate processes such as dissolution, percolation, volatilization, adsorption, biodegradation, hydrolysis, and complexation as well.

Scientists use results of the SESOIL model to describe

- Concentrations of a chemical in the water, the soil, and the air phases of a user-described soil column (consisting of up to four layers of soil with each layer containing up to ten sublayers of soil);
- Volatilization of a chemical at the soil surface;
- Transport of a chemical in washload due to runoff and erosion at the soil surface; and
- Transport of a chemical to groundwater.

SESOIL can provide results on a monthly basis for up to 999 years of computer simulation time.

Use of SESOIL provides students an opportunity to use technology that is currently used by scientists to understand and describe real-world environmental issues. It provides the student an opportunity to access and use information found in databases containing real-world environmental information. Further information on SESOIL can be found at *www.scisoftware.com/products/sesoil_overview/*.

Finding Online Data, Model Examples, and Resources in Earth Science

Data and resources are available from a number of sources on the Internet. These data may be used to assemble models of phenomena that are difficult to study directly in the typical classroom. Because databases change as web sites change, it is difficult to identify sources with certainty. One web site with a number of Earth science resource links—the Earth System Science Online (*www.usra.edu/esse/essonline*)—is sponsored

by the Universities Space Research Association.

The Online index contains two sections of particular interest to us as Earth science teachers: one contains ESS Resources, and the other, ESS Issues. Resources include links to sites such as the Global Change Data Center, Landsat Images of the U.S., Sea Surface Temperature Reports, EPA Envirofacts Warehouse, and National Climate Data Center, among others. The Issues database contains links related to topics that include global warming, El Niño, La Niña, sustainability, land use, agriculture, forests, climate change, ocean and coastal resources, and a number of other topics as well. Government agencies such as NASA, EPA, and USGS also have databases.

Limits to the Use of Computer Technologies to Teach Science

It is easy to argue that we should include instruction using computer-based modeling, probes, and technologically accessible databases merely because such instruction expands student awareness of these tools and because these resources provide for improved learning. However, poor planning may result in a considerable sacrifice of time and effort. (The time constraints to using computer technologies are especially significant in middle school. Teachers must "cover" so much material for the testing programs that there is barely time for a few hands-on labs.)

First, we should keep in mind that some software on the market is of questionable value. In particular, overuse of tutorial software may simply reinforce the idea that science is about learning facts and relationships, rather than about building meaningful models through engagement and exploration. Tutorial software has its uses, but it should not be the primary use of technology in science education.

Second, we should be aware that searching for databases and information on the Internet can be a time-consuming process that may turn up only modest results. Some databases are too complex, and others do not provide the particular kind of data needed to address a particular problem. We must know something of the data available to students before engaging them in activities that might use those data. At the same time, we should keep in mind that engagement with data—finding important patterns and relationships—can be as important as, possibly more important than, constructing a complete model.

Prior to using software programs such as STELLA to develop and explore systems, we must make sure that students understand the systems and concepts they are expected to model. We can use traditional means to achieve this objective. Diagrams of the water cycle, for example, can provide a basis for constructing a dynamic representation of this system. Once a system is set up and running, changing the values of variables—or adding new variables—will result in new outcomes for students to explore. Important here is to develop notions of systems boundaries, subsystems, feed-

back loops, growth curves, graphical relationships, and similar concepts related to system behaviors, as well as outcomes.

Students building models using Model-It, STELLA, or other similar programs may become frustrated because they are not working with real-world data. Lacking access to values that actually represent the variables in the model, students must create values they believe (but are not at all sure) may represent actuality. Because much of what we do in education focuses on gaining factual knowledge, students may be inclined to construct models with this goal in mind, and as a result, will be less likely to focus on exploring the behavior of the model system itself. We need to remind them—and ourselves—that investigating the behavior of models is an important reason for dynamic modeling, perhaps even more important than representing any one system accurately. In other words, if data are available, it is certainly permissible to use them, but it is still possible to explore the behavior of systems without such data.

Teaching with Computer Models and Technology

Technology greatly facilitates modeling in science by making data collection and processing easier and quicker, by opening up access to databases and resources that can be used to construct models, and by making static and dynamic models easier to build—and, in the latter case, easier to test. Technology does not change the basic processes of education and learning, but it can greatly enhance these processes. Because technology has become so embedded in the practices of science as a profession, students in our pre-college science classes should have as much experience using an array of technologies as can be provided. The following suggestions should govern such use.

First, *technological applications should reinforce the idea that science is a model-building activity and should facilitate the development and testing of such models.* This suggestion is a reiteration of the argument that the science curriculum should not be taught as the transmission of knowledge or a collection of processes, but should be regarded as the active development of conceptual models. In science itself, scientists adhere to certain rules to construct models; they communicate—through representational models—with others for analysis and subsequent acceptance, rejection, or modification.

We should not consider the use of technological tools as something apart from this process of knowledge generation, but as an integral part of it. In other words, the notion that models are tools that are sometimes constructed and used to represent reality (and sometimes not) must be replaced with the understanding that models are integral to any kind of thinking. The building of representational models is a requirement of thinking and of communicating beyond the confines of our individual mental models.

Expressed models are required in some form whenever knowledge is shared, while mental model building is required for all learning. Moreover, just as we may consider the telescope an artificial extension of the eye, so we may think of all technological

tools as extensions of some human organ—brain, ear, eye, arm, and so forth.

Second, *students should build models, with or without technology, that are appropriate for their levels of experience and ability.* This may appear obvious, but often students have difficulty tracking relationships, even in simple models. Furthermore, students may be unfamiliar with the process of model building. By framing science in the context of model building, students will have progressive experiences building on this theme. Providing experiences that are too advanced for students, however, will lead to discouragement. Using a program like STELLA with students who do not have a sufficient background in mathematics would be worse than providing no modeling at all.

Third, *students should be able to identify components and build static models before progressing to dynamic models.* It is generally difficult to build dynamic models on a computer without understanding the process of building static models on paper. The components of scientific models can be learned— and learned well— by analyzing static models such as laboratory reports, field studies, and so forth. Concepts such as dependence, independence, variable, constant, causal relationship, feedback, and correlation should be meaningful parts of students' conceptual framework before they engage in dynamic modeling. By framing inquiry as a process of model building, we can help students build a framework for understanding science as the study of dynamic systems. However, students should have much experience building models empirically before they approach the study of systems on the computer.

Fourth, *place emphasis on the process of model building as well as on its products.* Most of us continually struggle to find the proper balance between the relative emphases that should be placed on processes and products. Mindful that many standardized tests focus on content knowledge, we are naturally anxious that our students are prepared to do well on these examinations. However, in regard to meeting national goals, there are still many ways to emphasize processes without losing sight of the products. This may entail some tradeoffs, since time is limited. Use of technology has the potential to speed up some tasks—such as the collection and processing of data—and in and of itself can convey something about the nature of science.

Fifth, *be explicit in teaching how technology relates to science and how the specific technology used in an activity models its use in professional science.* A considerable body of research indicates that students do not necessarily apply what they are doing in the science classroom to larger contexts. As effective teachers, we should constantly try to link activities in the classroom to the larger context of professional science and technology. Some examples follow:

- After students use probeware to measure and record pH in water samples in the classroom, we can lead a discussion about how professional scientists use similar instrumentation to construct their scientific models.

- Although Model-It and other computer programs may be instructive in exploring

a phenomenon such as global warming, most students will need direction and/or direct instruction.

- Without explicit discussion, many students may not see the relevance of computer modeling to their own lives and interests.

At every opportunity, we should reach out to develop a continuum or framework for instruction that relates modeling activities in the classroom to the context of science.

Sixth, *over time, build up the general notion of systems*. For the most part, traditional education has not emphasized the study of systems. Students have tended to focus on learning discrete parts, but not on relating these parts to each other. Concept mapping is one way of dealing with this problem, but it may not be sufficient in and of itself for the depth of comprehension needed to understand important elements and relationships in more complex systems. Concepts such as correlation, causation, negative and positive feedback loops, systems boundaries, subsystems, and similar ideas related to systems theory are seldom introduced directly during science instruction, yet they underlie much of our contemporary thinking in science, business, and engineering. Dynamic modeling is based on systems theory. However, we must remember that understanding static models is desirable before we introduce dynamic systems modeling.

Seventh, *provide for exploration as well as direct instruction*. Exploration is, of course, the basis for true inquiry. And although open inquiry is difficult due to time constraints, it is essential for full development of science literacy. The best modeling programs are those that allow students to explore the interrelationships of different variables by manipulating their values and assessing subsequent outcomes. In these programs, the focus is on determining how manipulating variables can affect outcomes rather than on developing a specific concept—though concept development is certainly desirable. However, explorations need to be followed up with discussions leading to principles, hypotheses, or insights that make the activity worthwhile—including insights about the use of computer modeling as a tool for research and understanding. Without such discussions and analysis, some students may view the time spent creating computer models as wasted.

Eighth, e*ngage students in computer modeling in pairs or small groups*. It is generally true that individual students may profitably use computer software for tutorials. But computer modeling, like most inquiry, is usually best done in small groups, followed by large group sharing and discussion. Successful use of most computer modeling programs requires thoughtful analysis of the system being modeled, followed by construction and inputs that are best accomplished through team effort. Designing a model—as is done in Model-It and STELLA—requires the same kind of creative thinking as designing an empirical investigation; the mathematics required for some models, such as STELLA, is often easier when more than one person is involved in the design. Although students with a particular interest in and adeptness with computer

modeling may undertake individual projects, generally, small group work and sharing should be encouraged.

Last, *always relate science inquiry, whether using technology or not, to the way we think and learn*. Learning is basically a process of concept formation, hence the creation of conceptual models.

- Conceptual models do not have physical existence—they are theoretical entities to which we give meaning.
- Conceptual models represent ideas or entities in our minds that we may share as representational models.
- Conceptual models in our minds are mental models.

These three types of models—conceptual, representational, and mental—together constitute the foundational elements of thinking and learning. Regardless of the activities in which students engage, we must make every effort to ground science in the construction of models.

Conclusion

Computers enhance our ability to develop and test models of systems. The speed with which computers work and the number of variables they handle allow them to test the effect of many combinations of variables within a system. In addition, computers may be used to produce dynamic visual models to display results. It is not an overstatement to say that computers are essential to scientists who map factors and relationships in systems related to global climate change, earthquake dynamics, and weather forecasting.

Ideas and activities that involve modeling with computers should be a key part of the science curriculum. Although computer models have the limitations of any models, they are especially useful for modeling events that occur too rapidly or too slowly for us to observe, or events that are too dangerous or too expensive to do in a laboratory. Computer models can simulate events, modify variables, repeat events under various circumstances, and in general give students feedback on their experimental work.

CHAPTER 6

Inquiry and Model Building

Overview

Inquiry is a defining goal of contemporary science education and is prominently highlighted in the National Science Education Standards. Inquiry activities are not the same as "traditional" or validation laboratory activities, which often lack a serious inquiry component and are intended to confirm ideas that have already been presented. Inquiry involves student-centered exploration and problem-solving.

The process of inquiry, of course, is not limited to science. It occurs in a variety of human intellectual endeavors—for example, police inquire into a crime, a historian inquires into a particular event, or we inquire into matters concerning us. Each type of inquiry tends to be distinguished by its own rules governing what are acceptable assumptions, procedures, and outcomes. Scientific inquiry tends to be distinguished from other kinds of inquiry by its systematic approach and by the criteria for acceptable data and analyses, but there are overlaps.

THE NATURE OF INQUIRY

According to the National Science Education Standards, inquiry includes observing, posing questions, researching books and other sources of information to see what is already known; planning investigations; comparing what is already known with experimental evidence; using tools to gather, analyze, and interpret data; suggesting answers, explanations, and predictions; and communicating findings.

Even before John Dewey, educators championed the idea that science instruction should include practical laboratory activities. The arguments for lab work have varied, but include reasons such as to

- model professional science activity and raise interest in science as a career;
- develop scientific laboratory and problem-solving skills;
- facilitate achievement of kinesthetic and concrete learners;
- personalize and improve attitudes toward science;
- increase depth of understanding of important concepts; and
- raise achievement on standardized tests.

PROCESS SKILLS IN THE 1960S

In the 1960s, a renewed emphasis on developing scientific process skills began with a series of national curriculum packages, the so-called alphabet curricula. These included the Biological Sciences Curriculum Studies (BSCS) blue, green, and yellow series, and the Intermediate Sciences Curriculum Study (ISCS) series. The alphabet curricula met with resistance from some teachers, who felt them to be too time-consuming. The net effect, these teachers predicted, would be a watered-down curriculum and a loss of conceptual learning. This prediction proved not to be the case, as several meta-analytical studies carried out in the 1980s illustrated. On average, students involved in the process-oriented alphabet curricula gained as much in knowledge as did their counterparts in more traditional classrooms—and more in affective variables such as attitude and the ability to use process skills.

Research shows that inquiry, when well used, is effective for teaching scientific concepts in depth, stimulating student interest, and developing critical thinking skills. Our current notions of inquiry in the classroom go beyond just the use of a particular set of process skills, however. Instead, they focus on the creation of a state of mind that values systematic problem-solving based on the tenets, principles, and processes of science. Over the years, the idea that there is only one "scientific method" has given way to the idea that there are many ways to approach a problem in science. What is most important is the outcome: Is the model the scientist has created and presented

consistent with our basic assumptions, current explanatory models, and standards of evidence?

Over the last century, we have witnessed a shift from a rather strict cause-and-effect view of the universe in which it was possible to discover and know truth, to a view of reality that is more relativistic and probabilistic. In this contemporary worldview, inquiry yields results that have a high probability of being predictive, but which cannot be absolutely verified. In this sense, the saying "The one absolute is that there are no absolutes" is true in science.

There is general agreement among science educators that to be of value, inquiry should lead to one or more of the following results for students:

- Better ability to use the processes of inquiry to solve real problems.
- Deeper understanding of concepts studied through inquiry.
- Greater understanding of how science relates to one's personal life.
- More interest and positive attitudes toward science.

Recent research indicates that inquiry activities in the classroom do not result in better understanding of the nature and practice of science in and of themselves. Instead, inquiry must be situated within a larger framework of human beliefs and practices that define science as an enterprise. The purpose of this chapter is to propose how we can better contextualize inquiry as part of this larger set of scientific practices.

The Role of Scientists as Model Builders

If we ask a typical student what scientists do, chances are that he or she will reply simply that they do research. Draw-a-Scientist tests reveal an image of scientists held by many students of a male, dressed in a lab coat, often with thick glasses and frizzy wild hair, surrounded by smoking laboratory apparatus. While it may be that students draw such images tongue-in-cheek, the notion that scientists are unique types of beings who create weird and sometimes frightening things appears to be part of our cultural stereotype. Even apart from this, many students do not have a well-defined image of what scientists actually do. "They experiment" and "They do research" are common responses to the question: "What do scientists do?"

These responses do not define what scientists do, any more than "they catch and hit baseballs" defines what baseball players do. The practice of science is more than just experimenting. Missing from these responses is a sense of the overall goals and processes of science from start to finish. To flesh out this definition, let's take a look at model building as a defining theme for the practice of science.

Scientists are professional learners: As such they are professional model builders as well. The role of scientists in society is to construct consistent and predictive models grounded in empirical experience. We find scientific models useful because they define the perceptible world in a predictive way so that we can act effectively within it.

Science, then, has the goal of building and communicating a particular kind of model. It is not simply a process of inquiry. Inquiry is important, but so too is communication: construction of the final model from the data to present to others. Other aspects of science are important as well, including debates over the acceptability of various models, and the development of theoretical models from various working models proposed by researchers.

Teaching Implications of Using Models to Frame Inquiry

There are many reasons for students to undertake inquiries in the larger context of scientific model building. Model building provides students with a broader framework for understanding and practicing science. The purpose of a lab activity, for example, is not just to inquire; rather, it is to create an intellectually satisfying and persuasive model of the target system based upon observation and inference, and to present and defend that model to others. This process better reflects the spirit of science than does inquiry in isolation.

INQUIRY SKILLS

Following are some integrated skills of inquiry. At each step, learners are building, evaluating, and communicating their mental models of the science concept being studied. By the sixth process skill, students should understand enough about the science concept to be able to assemble their own model of it.

1. Controlling variables — being able to identify variables that can affect an experimental outcome, keeping most constant while manipulating only the independent variable. Example: realizing through past experiences that the amount of light and water need to be controlled when testing to see how the addition of organic matter affects the growth of beans.
2. Defining operationally — stating how to measure a variable in an experiment. Example: stating that bean growth will be measured in centimeters per week.
3. Formulating hypotheses — stating the expected outcome of an experiment. Example: predicting that the greater the amount of organic matter

added to the soil, the greater the bean growth will be.

4. Experimenting — being able to conduct an experiment, including asking an appropriate question, stating a hypothesis, identifying and controlling variables, operationally defining those variables, designing a "fair" experiment, conducting the experiment, and interpreting the results of the experiment. Example: going through the entire process of conducting the experiment on the effect of organic matter on the growth of bean plants.
5. Interpreting data — organizing data and drawing conclusions from it. Example: recording data from the experiment on bean growth in a data table and forming a conclusion that relates trends in the data to variables.
6. Formulating models — creating a computer-based or physical model of a process or event. Example: developing a model to show how the addition of different amounts of organic matter affects the growth of beans

Students in the classroom are often asked to complete labs lacking a context that gives the process of exploration and inquiry a broad and more satisfying meaning. They may be required to fill in tables with data and provide "right" answers to preset questions. Science is thus reduced to the completion of particular tasks. Model building gives inquiry more form and substance. Consider the following simple example of instructions by a teacher:

> "Today we are going to use string and tacks to trace out the geometric figure formed when an object moves around two focal points at the center of the figure. You should then answer the questions about the figure that are on the back of your instruction sheet."

These instructions require students to create a figure and answer questions. There is no sense of how this activity fits into the broad context of science. Now compare this to the following instructions for a guided inquiry:

> "Scientists often create concrete simulation models in order to understand what they observe and then present their ideas, their mental models, to others. Your goal today is to create a concrete diagrammatic model that represents the trace of an object such as a planet moving on a path around two focal points. Once you have finished your diagram, you should add to your model by providing a verbal narrative explaining your diagram, using the questions of the back of the page as guides. Be ready to present your complete model to others and to defend it."

The difference between these instructions is not trivial. In the first case, students are not provided with a context for what they are doing. The task stands alone, unrelated to science or to the processes scientists use in their work. In the second case, the task is related to what scientists often do, and is contextualized as a model-building process. In addition, the teacher expects a definite product—a model.

Inquiring with Pictorials

Diagrams, drawings, and pictures are essential components of teaching. We use these models to illustrate and illuminate our ideas. Issues of the journal *Science* and *Nature* are laden with diagrams, pictorials, graphs, and data tables. The typical textbook also contains many pictures and diagrams—and for a reason. Remember that human conceptual models have two essential components: images and propositions.

Propositional thinking might be thought of as a later evolutionary development. Other animals do not appear to have much, if any, capacity for it. Humans may be less adept at propositional thinking than at thinking through images, which may help to explain why mathematics is so difficult for so many students. The axiom that one picture is worth a thousand words is a restatement of the idea that "imagination" rules human thinking.

This being the case, it makes sense that image-models should figure prominently in teaching, whether in science or in any other field. If images play a major role in our thinking, thinking through images ought to play a major role in our pedagogy. Perhaps one of the reasons lab activity is so important, and has such an impact on learning, is that it consists largely of image building. (Remember that "image" in this sense is not just visual, but consists of all sensory inputs that are organized into holistic memory constructs). A good laboratory activity involves image building in a number of sensory dimensions. Ask science teachers what they remember from their college days and chances are good that the memories will be of lab or field activities, not hours in the lecture hall.

If you ask colleagues to talk through their thinking and to describe the solar system, it is likely that the first thing they will recall is a picture or diagram of the solar system—one similar to that commonly found in science texts. Ask others to describe a volcano and more often than not, they will come up with a mental image of a picture or diagram they have seen of a volcano. (If the interviewees have visited a volcano, of course, that may be their dominant image.). One way or another, people tend to rely upon images to frame their thinking about a particular subject.

This being the case, why do we spend so little time using the diagrams on charts and in texts to help students build accurate mental models of what we want them to learn? Diagrams and pictorials are potentially powerful tools for learning concepts, and for analyzing them critically as well. However, these visuals have their shortcomings. As

models, all diagrams and pictorials are incomplete representations. Photographs are close representations, but they lack depth and detail. Diagrams may suffer distortion in order to show the essential characteristics of the system being portrayed.

An excellent and almost classical example of distortions is the typical diagram of the planetary relationships found in many textbooks. What the diagram actually shows is the order of the planets in relation to the Sun. It is virtually impossible to portray size and distance relationships accurately on the pages of a textbook, because the dimensions of the pages do not allow it. However, the image that students take away from the textbook is of the Sun and planets as they are portrayed on the page. We generally do not point this problem out to our students, nor lead them to question it.

Since the world of science and of experience is often narrowed down to the images in films and texts we provide in our classes, we want students to study these images, not only to acquire a concept, but also to detect what is not in the image.

For example, let's consider the study of maps. A topographic map is a pictorial model. What can one *imagine* the landscape depicted to be like? What information is present in and missing from the model? If we were to stand at a particular point on the map and face south, what would the landscape be like? Could we draw what we would see? This kind of more detailed analysis—filling in the gaps in the model—will lead students to a critical regard for the model they are studying, and will give them a richer understanding of the strengths and weaknesses of the illustrative models we use to represent natural systems.

Even when models accurately illustrate a phenomenon, it may be useful for students to consider that which is missing. Consider a typical cross-section of the layers of the Earth to illustrate the origin of artesian wells. At best, this is a model of what we believe exists. After all, no one has cut away such a large section of Earth and actually witnessed the flow of water as it is described in the diagram. That the diagram is accurate is largely a matter of trust. It explains the phenomena and permits accurate predictions based on this explanation. But what is actually happening within this system? Are there flaws in the layers of rock? Are rock layers impermeable or permeable? What would these rock layers actually look like if we were to see them? What is not shown: the rate of movement of water, the actual movement of the water, the weight and solidity of the rocks. None of these important features of the natural system can be detected in a diagram.

Thus, to understand models like the artesian well system, students must be able to translate the model conceptually, and form an image of the real system in their own minds. In other words, in contrast to the translation of the real world into a diagram for the textbook, we must think now about inquiring into the way the diagram can back-translate into a real-world conceptual image.

In a recent issue of a scientific journal, there is a very nice picture of a water droplet. The drop has struck the surface and the water has come back together to form a pillar of

liquid. Meanwhile the water around the strike zone is spreading out in ripples. This is a familiar picture to most of us, yet what is happening here? How can we explain it? Why has the pillar formed to defy gravity, and what is happening in those ripples?

Figure 6.1

Water droplet striking a liquid surface

Photo by Photodisc

Students may well be able to explain energy transfer, cohesion, and adhesion in the usual terms required by standardized tests, but as they look at this diagram, can they explain how these ideas can be applied to the image before them? Here is a photographic record, a type of model, before them. What can they see? Clearly, the picture shows a state of being frozen in the moment. But much is not there: the motion of the waves, the rise and fall of the pillar of water, the refraction of light—the overall display of energy dynamics implicit in this moment are frozen in time, but are not available to those who study this model.

However, the scientific pictorial model has its uses. We can take the time to consider what is happening. Without the photograph, the action might take place too quickly for analysis. Indeed, that is the purpose of scientific models in many cases: to show a state of matter or action as it is frozen in time. Basically, the model is useful only if it can be translated as the author originally intended, which is that we should see and understand a dynamic reality from the model. Unfortunately, this type of perceiving seldom happens in the classroom. Models are learned as models, and the characteristics of these models become part of students' conceptual model of the world.

To back-translate, students must use their imaginations to build a conceptual model that has the elements that are missing in the pictorial model. Students may build a propositional model (write a short description) reinterpreting the model so that it exists in their own imaginations as it might exist in reality—adding smell, sound, and vibrations to the photograph of the erupting volcano, for example. Similarly, to back-translate a natural system, students might take a (real or imagined) trip underground to examine the system of rock and water that gives rise to an artesian spring.

Such translations should not be regarded as trivial. The ability of scientists to imagine their world is probably one of the most important elements in the creative practice of science. Because proof and evidence are required to confirm an idea, these factors are sometimes given undue emphasis in discussing science processes. But the first step is actually to imagine what is there in the first place. Scientists do not have diagrams to imagine from, but it stands to reason that the ability to imagine can be developed by back-translating and interpreting existing pictorial models.

Investigating a Problem

This section demonstrates how to guide a class through a scientific investigation. Ask students to develop a guiding question, such as the following: *What is the effect of soil type, ground cover type, and rainfall amount on the quantity of water runoff?* Here is a brief overview of the different activities students might do to investigate the question:

1. Have students identify the problem in the context of developing a model. This step involves describing the driving question, deciding on the factors involved in the question, stating a hypothesis, composing a general plan for the investigation, and recording research notes. The problem in this case can be stated: To develop a model of how soil type, ground cover, and amount of rainfall influence the quantity of water runoff.

Figure 6.2

Statements for Research within a Models Framework
I would like to build a model describing changes in speed on motion of a pendulum with changes in its length.
I would like to build a model of *E. coli* populations in Middleburg Creek from January to June.
I would like to build a model showing changes in the position of the Sun relative to the horizon from March 1 to June 1.
I would like to build a model of the accuracy of weather predictions one, three, and five days out.
I would like to build a model of soil permeability to water under different land use conditions.

2. Ask students to describe how to construct the model. For example, "I will use two different soil types (clay and loam) and two different covers (leaves and grass) in large stream troughs. I will dry each to approximately the same level, then use a watering can to add five liters of water to each. I will collect and measure the runoff and present the results as bar graphs."
3. If students have expectations, they should state them and give the reason for them.
4. Have students run the experiment and collect the data.
5. Students should then process the data and find and interpret patterns they observe.
6. Finally, ask students to assemble the model in the format of a research report. They should build this expressed model step by step, describing how they think the system behaved and why, in relation both to their expectations and to other things they know about similar systems. Encourage students to generalize their findings to a larger target and assess how confident they feel that their model represents the target.

   Let's see how these ideas play out in a classroom.

Example of an Inquiry Activity: Mrs. Watson's Class

Mrs. Watson is planning an inquiry activity for her tenth grade Earth science class. In addition to the content and processes of science she would like to convey, she also wants to develop the notion of scientific models. The unit the class is studying is on atmospheric science, and Mrs. Watson is focusing on the concept of convection. "What I want to do," she explains to Mr. Johnson, a fellow Earth science teacher, "is to involve my students in constructing a scientific model of convection through inquiry. I'm going to have them observe several convection systems and see if they can pull their observations together to create descriptive and explanatory models of the events."

Mr. Johnson nods.

"What's the difference between that and just doing inquiry?" he asks.

"I think it's the way we phrase the explanation," Mrs. Watson replies. "It's the focus on model building. It takes us one step beyond inquiry by giving the students a more holistic idea of what science is about. Right now, a lot of them seem to think scientists just do labs, like they do in science class. They don't understand that the whole process of science involves more than just research. You have to pull together your ideas and integrate them with what is already known. You have to make sense of what you see and report it clearly. That is, you have to build a model of the thing you observe."

"Sounds interesting," Mr. Johnson admits. "Let me know how it works out."

When the class meets the following day, Mrs. Watson distributes the materials and instructions for an activity that will allow her students to observe a convection phenomenon in a glass bread pan filled with water. She asks students to observe what happens as they carry out her instructions and to record what they think are significant changes in the system.

"What we want to do first of all," she tells them, "is to construct a descriptive model of changes in the system you are observing. A descriptive model is the fundamental building block for other models we will build later. The building blocks of your descriptive model are the data you will record. What are some of the ways you might record the data for this set of observations?"

The class is silent for several moments before Trent speaks up. "Our notes? You know, what we write?" he says.

"Good. Anything else?"

"Drawings we might make," Alice interjects.

"Okay." Mrs. Watson waits expectantly. The class is silent, so she goes on.

"All right, your notes and drawings are two kinds of data you might use to construct your model. You might find other ways once you get started. What's important is that you clearly show the changes you see taking place. Scientists have to do the same thing when they are confronted with something that is new or unusual. They are like detectives. The descriptive model comes first, then the explanation."

The students begin work. They set up the activity as directed in the materials Mrs. Watson has provided. She circulates among the work groups, asking and answering questions, being careful not to provide too much information that would destroy the inquiry process. She encourages groups to repeat the activity several times to be sure their observations are consistent. When all the groups have completed their work, she brings them back together and asks them to share their observations with each other.

In this case, all six work groups appear to have come up with similar results. Mrs. Watson invites Bill's group to record the observations of each group on the board. They do this without analyzing the observations, until all have been recorded.

"The descriptive model we are building needs to be consistent for all six groups," she explains. "Does anyone know why this is important?"

Trevor raises his hand. "If two or three were different," he says, "you wouldn't know which one to trust."

"In science, descriptive models have to be trustworthy, that's right," Mrs. Watson replies. "We call this reliability. We can build a reliable model either by doing the same observations repeatedly or by having a number of people do the same observations. Do you think the observations among the groups are reliable?"

The students nod their assent.

"Then what we need to do next is for each group to use these observations to construct a descriptive model of what you saw."

After a time, Mrs. Watson again addresses the class. She asks Lori's group to share its descriptive model. As Tisha holds up a diagram the group members have made, Lori reads their written description. "Once the hot plate was good and warm and the system had been running for several minutes, we dropped the food coloring in the middle. Some stayed on top and some went to the bottom. The color on top started to move—most moved to the cool end of the pan. The color on the bottom moved—most toward the warm end. When the color got to the warm end it spread out and rose to the top. In the cool end, the color kind of spread out, but it looked like it sank and moved back toward the warm end. Finally the color was all spread out in the pan. We did it again with clean water, but this time we put the drop closer to the cool end. It looked like most of the color moved toward the warm end along the bottom, then rose to the top over the hot plate and moved back."

"Lori, do you and your group think you've created a good model of what you saw?"

Lori and the others nod. "I think so," she says.

"Let's review what happened here," Mrs. Watson says. "When you observed what was going on in the pan, you were processing information and creating a model of it in your mind. What kind of model is that?"

"A mental model," Lori replies.

"And from the mental model, Joan, we created the model that you drew and wrote down and shared with the class. What kind of model is that?"

"A scientific model? A descriptive model?" Joan replies uncertainly.

"It is a descriptive model, yes. We probably wouldn't think of it as scientific yet. But what is the term we agreed to earlier for a model that represents our mental models to others?"

"An expressed model," says Trevor.

"Do you people in the other groups agree with this expressed model Lori's group has created?" Mrs. Watson asks, looking around at the class. They nod in agreement.

"Then the next thing to do is to observe another event and see if it is related to the expressed model you've just created. As you know, new observations sometimes complement a model we accept, and other times they contradict it. Let's see what happens here. We'll have to do this as a demonstration because we need to use the hood. Kevin, could I ask your group to run this activity?"

In short order, Kevin's group sets up the activity under the hood. The lights in the room are turned out and students observe what happens to the chalk dust in the projector beam when the hot plate is cool and when it is warm. As instructed, they do not make inferences but speak to each other only to compare their observations. When they are satisfied with their observations, they return to their groups. Group members compare notes, then complete an expressed model depicting what they have seen.

"So what does your new model represent?" Mrs. Watson asks. "Does it complement or contradict your previous model?"

Claudia raises her hand. "We thought it complemented it. The chalk dust rose above the warm hot plate but not the cold one. It was just like the food coloring when it rose in the warm end of the pan and not in the cold end."

"So this extended your descriptive model?"

"I think so."

The others in the class agreed.

"We're almost out of time today. Tomorrow we're going to combine your observations of both systems and create a general descriptive model that would include them together. We're also going to develop an explanatory model of this phenomenon we call convection."

The next day, the class settles in their work groups and Mrs. Watson begins. "Yesterday we built descriptive models of two systems and related them to each other. Let's think now about why the changes we saw might have taken place. When we deal with the questions of 'why'—explanations—we are creating hypothetical models. Why do you think the food coloring and the chalk dust rose over the hot plate?"

"It has to have something to do with the heat," Chuck says.

"When the food coloring and chalk dust got warm, they went up," Tisha interjects.

"What else could be rising?" Mrs. Watson asks.

"The water or the air?" Claudia says hesitantly.

Mrs. Watson smiles and waits.

"I mean, the food coloring is in the water. It could be the water or the water and the food coloring together that go up. Same thing with the air and chalk dust. Like sparks over a campfire," Claudia continues.

"So why would warm air—let's just focus on the air—rise?"

Greg pipes up. "Well, we did an experiment with a balloon in eighth grade and filled it with helium and let it go, to see how much weight it would carry. The helium floated because it didn't weigh as much as the same amount of air."

"So something will float on something else if it weighs less?"

"Yeah, like a cork on water or something. It'll float, but a piece of steel the same size will sink. We did that in eighth grade too," Alan says.

Mrs. Watson considers. "Let's test that idea by building another descriptive model. This time we'll study what happens when we mix warm and cold water together. Over here I have some refrigerated water and some very warm water. The warm water is colored. We're going to mix them carefully and observe how they behave in relationship to each other. I have the instructions for the activity here. Work in your groups, do the activity, and construct your descriptive models. Then we'll get together and discuss the results."

The students follow Mrs. Watson's directions. In a short time, they have determined that very warm water floats on top of colder water. While it is now clear that their models have expanded to include elements of explanation, the relationship to

density is not established. For the sake of time, Mrs. Watson chooses to build on their prior knowledge rather than to present them with density activities.

"Did you study the concept of density?" she asks.

The students nod hesitantly.

"What is density?"

The students struggle to recall the meaning of density from earlier classes. At last they agree that it is the amount of mass in a given volume, and that the explanation for floating objects is that they are less dense than the same amount of the substance in which they are floating.

"Could this model be tied to the descriptive model you just created for the way different temperatures of water interact? Your next task is to work in your groups and tie this model together with the models you created yesterday. See if you can develop a hypothetical model that would explain the phenomena you've observed."

The students return to their work groups and discuss the latest task. With time, they arrive at the conclusion that one explanation of what they have observed is that warm water or warm air is less dense than cool water or air. As both the air and water are warmed by the hot plate, they become less dense; they then rise and carry the chalk dust or food coloring with them.

"Is this the only possible explanation for what we observed?" Mrs. Watson asks.

"Maybe not the only one," Claudia says, "but it makes sense based on what we saw."

"Why might we consider your explanatory model to be a scientific one?"

Pete raises his hand. "Because it's based on what we saw," he says.

Mrs. Watson smiles. "But what if we said it was the water spirits that made the hot water different than the cold water. That would explain what we saw too, wouldn't it?"

"I suppose," Pete admits, "but you don't explain things in science by talking about spirits."

"The density model explains what we saw okay," Dee follows up. "We don't need spirits to explain what we saw."

Analysis of the Example

By infusing the idea of modeling into science and scientific inquiry, we can provide greater context for the "doing" of science than is the case without such infusion. In the example of Mrs. Watson and her class, she is able to make a number of points about science, including the need to synergize multiple models to create new models, and the relationship between individual mental models and the expressed models we create to portray them. She is also able to make points about the nature of scientific models, as she does in the last several lines.

Perhaps most important is that Mrs. Watson leads students to conceptualize their activities and actions in the science classroom in a different way than when they are just given labs to do. Each phase of the learning process has a task focus: to build a

model that is descriptive or explanatory. This task focus accurately reflects the task focus of the practicing scientist. The phenomena are not the investigations, but are opportunities for investigation and model building. This distinction is important, as too often we associate science with certain phenomena (oceans, weather, plants, and animals, for example) that should be part of the conceptual world of any intelligent individual, regardless of whether or not that person is of a scientific mind.

People in professions or jobs outside of science may not conceptualize their role as that of doing research or conducting scientific investigations, but they can appreciate the act of building a scientific model of a thing, and perhaps more importantly, of distinguishing between scientific and nonscientific models.

Conclusion

The message of this chapter is the importance of contextualizing inquiry so students understand that science is not an isolated set of inquiries, but that inquiry is set within a larger context of the creation of theoretical models. This thinking requires students to understand the concept of models in a broader sense than most people do—in a sense that is consistent with the way the concept of models has been presented in this book. Our brains model reality in the form of mental models, and we create expressed models, such as research reports, to show others these mental models—our ideas.

The chapter also stresses that we adopt a new vocabulary to better reflect the notion of science as model building. We talk about building a model of density, rather than doing a lab on it. We should learn to use the terms "hypothetical model" and "theoretical model" rather than hypothesis and theory. The mere use of these expressions reinforces the idea that we are building models, which in turn reinforces certain principles and assumptions underlying modern science.

Models and Teaching the Nature of Science

Overview

The description of learning as a model-building process—that is, as the process of building a mental conceptual model—is particularly advantageous when relating science to other fields of intellectual activity.

When we conceive of our thoughts as models, we can better understand and accept the limits of learning. In fact, our ability to distance ourselves from our conceptual creations is at the heart of critical thinking and scientific skepticism. Holding out our thoughts as models gives them objectivity: We can examine them with a critical eye, as an artist examines his or her creation, acknowledging that not everyone is going to create the same model, nor needs to. This recognition allows us to accept and consider how the models of others may be different from ours, and prevents us from being unnecessarily defensive of our own models.

We should also understand that knowledge—as it is expressed to us by others through literature, film, lecture, and so forth—consists of their models of reality, not reality in any absolute sense. This idea has been confirmed by eyewitness accounts of the same event, which frequently vary widely. We always have to decide which models to accept and which to reject.

We usually make these decisions based on models' agreement with those we already hold; in decision-making, we use individual models that determine what is and is not acceptable to us. Part of what we are trying to accomplish as science teachers is to assist students in building mental/conceptual models that will allow them to decide which models are scientifically valid and which are not.

Models and Pseudoscience

Millions of dollars, and in some case many lives, are lost each year because we cannot distinguish between scientific, pseudoscientific, and nonscientific models. The damages incurred range widely, and include lost individual savings, missed opportunities for proper medical treatment, squandered research funds, and wasted legal resources. In some cases, pseudoscience has set research programs back decades, as happened in Soviet Union in the first half of the twentieth century, when political reasons led to favoring the doctrine of acquired characteristics over Mendelian genetics.

Some pseudoscience occurs because of poor research methodologies, but most often it is the result of deliberate actions by individuals who

- think they are doing science, but are ignorant of principles of scientific inquiry and standards of evidence;
- are skeptical of "standard science" and are trying to argue a case for a phenomenon with weak evidence; or
- profit from pseudoscience by attracting a following to support their cause and/or organization.

Science teachers can help students become intelligent "consumers" of science and avoid the pitfalls of pseudoscience. To defend against pseudoscience and the acceptance of nonscientific claims as science, teachers must point out that scientists have certain tenets and standards of evidence. To summarize some of the more important tenets, scientific models must be

- Testable through observation and prediction. Tests of science are ultimately based on whether or not they can predict events or support inferences of past events based on observed phenomena.
- Falsifiable. Claims that cannot by their nature be shown to be false are not in the province of science.
- Consistent with established models. Inconsistent models require extraordinary evidence to be established in science.

THESES REGARDING THE NATURE OF SCIENCE

The following philosophical theses regarding the nature of science as process underlie much of the contemporary thinking about science, though they are not accepted by everyone.

- The material world exists, or acts as if it exists, independent of human thinking.
- Human knowledge is tentative because there is no way to know all possibilities.
- Scientists more often modify than replace existing ideas.
- Science does not deal with ideas that are not empirically testable.
- There are many methods for fruitful scientific inquiry, not just one.
- Science is distinguished from nonscience by standards of evidence accepted by scientists.
- That which is science is determined by the scientific community.
- Science is subjective, but ideally scientists try to avoid unwarranted biases.
- Overall, the direction of research is influenced by prevailing views of the society or culture.
- Ethical considerations have an impact on what is researched.

RESISTANCE TO SCIENTIFIC IDEAS

Throughout history, reputable scientists have knowingly or unknowingly contradicted the ideal of open-mindedness. In these instances, scientists resisted new scientific ideas, even though the new ideas met most of the criteria for "science." Often the causes of such resistance included

- faith in an existing model or set of assumptions;
- contradictory ideas about what constituted good scientific methods;
- religious ideas, social and cultural influences;
- threats to their professional standing;
- professional specialization—resistance to outsiders by an in-group; and
- societies and academies, schools of thinking, and seniority, especially those dominated by "old school" thinking.

Ways of Learning in Society

Most humans are constantly learning, either formally in schools and similar organizations, or informally by personal experience. As we learn, we create operational mental models that help us survive and achieve our goals. We test our models as they are needed, and over time we construct a framework of models that becomes the core of our personality.

Each model carries with it a certain truth value. This is the degree to which the model is aligned with the external world and meets our needs. Different knowledge models have different kinds of truth values. The degree to which we impart a truth value to any given knowledge model is always a matter of faith on our part. This is as true in science as it is in any other area of human knowledge. To accept science as absolute truth, one must accept the underlying assumption that truth lies only in the observable world.

Regardless of our core commitment to science or to some other way of knowing, each of us possesses many—sometimes contradictory—conceptual models. To meet the goal of scientific literacy called for in the National Science Education Standards,

our students must understand that science is only one way of knowing. It works to meet particular goals, but not all goals.

Before exploring the implications of these ideas, we should define what we mean by *knowledge* and *knowing operationally*. Knowledge includes all conceptual models we use to interpret and portray our worlds. For example, let's consider unicorns. Do they exist? Did they once exist? Of course, we have no evidence that unicorns existed, except in our imaginations. Are unicorns then part of our knowledge base? It would seem so. Once a concept becomes incorporated into our mental/conceptual models, it becomes part of our knowledge. We may not believe in the physical existence of unicorns, but these creatures have an existence in mythology. Like the concepts of love or force, shared knowledge is a label for a conceptual construct that does not have an independent existence in the observed world.

Issues in Teaching Science: Tom's Case

Tom is embroiled in a conflict with several of his students. Tom teaches in a rural midwestern school corporation. He has just introduced a unit of Earth's history and the fossil record. Though he has not mentioned the concept of evolution per se—he is aware that many in the community do not accept the theory of evolution—the students have nevertheless challenged the ideas Tom has presented.

Their position is that Tom should also discuss special creation and give it equal time. Tom has never had any doubts about the scientific explanation for the fossil record, and he attempts to dissuade the students. However, his arguments appear to have the effect of encouraging dissent. The students accuse Tom, a first-year teacher, of closed-mindedness and being unscientific, since he refuses to present alternative theories.

Paul, a biology teacher and Tom's friend, says that this difficulty is normal for students in this particular school corporation. Paul's solution has been to give minimal attention to evolution in his biology class, referring instead to changes over time. Tom disagrees with Paul's "solution" but is not sure how to handle the conflict. Tom clearly believes that special creation is a faith-based idea, and he does not want to be put in a position of attacking students' religious beliefs. At the same time, he is just as convinced of the truth of the scientific evolutionary theory and does not like the idea of, as he puts it, "politicizing science." What should Tom do?

Dealing with Issues: Types of Models

Any teacher who deals with issues is going to find times when challenges to science arise in the course of discussion. The theory of evolution is a prominent source of conflict, at least in the United States, but there may also be issues that arise with regard

to global climate change, depletion of natural resources, population growth, and the wisdom of spending on space travel. The sources of conflict may be rooted in misconceptions about what science is; there may also be conflicts about constructs such as social, cultural, economic, historic, religious, philosophical, or political models.

It is important, then, to make students aware of these differing models upon which they base their lives, and then to help students understand how these models interrelate to provide us with opinions and beliefs.

Special creation or intelligent design models of being are not aligned with—true with—the tenets and standards of evidence of science, but are fully consistent with some religious models. Since we cannot disprove these religious models based on their assumptions, arguments intended to do so will have little positive effect. What is important is that students understand the differing bases of the religious and scientific models, and do not see one as comparable to the other.

All of us have a right to our own beliefs. We cannot disprove either creation or intelligent design hypotheses—indeed, these hypotheses cannot be falsified either, so according to our criteria, they fall outside of science—but the scientific model for the theory of biological evolution is better aligned with observed evidence and the tenets and practices of science. What is important is not that students accept evolutionary theory but that they realize the difference between this model and various religious models.

Another example of how models mix underlies the effort to restore the Everglades ecosystems by engineering a return to a more natural state that will allow free water flow from Okeechobee to the Gulf of Mexico. In the planning stages, economic models have clearly become intermixed with scientific ones. If we are teaching ecosystems and use the proposed Everglades project as an example, we must be sure that students recognize the idea of competing, equally valid models so they can understand the controversy surrounding this project. Both sides may well accept the same models, but each assigns different priorities to one model or the other. In many controversies, competing world models are so different that true understanding can be achieved only through considerable dialogue, and in fact, may not even be possible at all.

This being said, let's look at some different kinds of models and dissect their basic goals and assumptions.

Scientific Models

Scientific models explain phenomena in empirical predictive terms. Explanations are based on data and assessments of reliability (repeatability) and validity. We assume in science that the universe works through a series of cause and effect relationships that are predictable. While random fluctuations occur, they are not guided by divine intervention; rather they result from factors that occur naturally in the mechanics of the universe.

Truth in science is determined by how well the models we create line up with observations and predictions. Science provides us with a material world that is relatively predictable and gives us a measure of control over it. Current science does not deal with ultimate causes or with questions that cannot be answered through prediction or observation.

Religious or Theological Models

Religious models are also called theological or spiritual models. The number and variety of religious models makes generalizing about them difficult. However, most major religious or theological models are concerned with our relationship(s) to one or more deities, and include an assumption of life after death.

Religious models are not data-based. Rather, they tend to be grounded in analogies, revelation, and authoritative scripture. These models emphasize faith and belief based on an internally consistent logic. The models do not generally require consistency with observed reality, they assume that the world is under control of one or more deities and can be altered by their will. Truth is determined by internal alignment with scripture or revelation. Observations or interpretations inconsistent with fundamental tenets of the religion may be rejected as illusory. The broad claims made by major religions tend not to be falsifiable, hence are not the subject of scientific study.

Until relatively recently, throughout the history of science, science and religion were pursued simultaneously. Newton, for example, accepted the idea that science was a way to know the mind of God. However, particularly during the eighteenth century, philosophers such as Immanuel Kant engineered separate domains for science and religion, and this separation has become the model for practice in the Western world.

Political Models

Political models are conceptual models dealing with power and control within a group. These models include statements of ideals, processes, assumptions, laws, principles, and examples, and they vary extensively in size and complexity. Broad conceptual political models include the ideas labeled *democracy, totalitarianism, fascism,* and *libertarianism*. Machiavelli's *The Prince* is a political model, as is the *Red Book* of Mao, Hitler's *Mein Kampf*, England's Magna Carta, and the U.S. Constitution. At a more local level, town charters and articles of incorporation are also political models.

None of these models is based on the same goals, assumptions, and principles as either scientific or religious models. Rather, political models usually evolve through experience and laws, and are shaped by values and ideals, economic needs, religious doctrines, and many other factors.

Political models are true if they achieve the goals of those in power, which may

be the broad citizenry in a democracy, or a select few rulers in an oligarchy or totalitarian state. As such, these models guide both those in power and those who are powerless as to their rights and expected behaviors. Political models may meet needs ranging from basic safety and security to idealistic world-shaping. They are power constructs that make governance and social interactions in large groups possible. Political models compete against each other in a nonlinear selection process that shapes societies and cultures.

Fictional Literary Models

Literary models are the stories we tell, whether fiction or nonfiction. Fictional literary models include such constructs as novels, short stories, scripts, and poems. They are stories invented by the author, but mirror life well enough to retain plausibility.

The purpose of a literary model is to create an imagined world. This mental model is then expressed in the propositions that make up the lines of the manuscript. As the reader reads, he or she constructs his or her own mental model from these symbols.

The truth in a literary work is its plausibility relative to the expectations of the reader. When we say a story doesn't "ring true," we mean it seems unrealistic in the context the author has provided. Most of us probably imagine we are enjoying the story in the same way the author envisioned it, but there is no way to affirm this. Chances are excellent that no two mental interpretations of a novel are very much alike.

The literary models that infuse a culture are very powerful in terms of their impact on individual and group beliefs and behaviors. These models reflect a mixture of many types of knowledge. From Shakespeare's plays to Stephen King's mysteries, stories are constructions that we read because they excite and often instruct us. They extend our ranges of experience and are often important to our sense of self, our sense of possibilities, and even to the choices we make. More than one engineer at NASA has cited Stanley Kubrick's film *2001: A Space Odyssey* and the television series *Star Trek* as positive influences on their decisions to pursue their careers.

Importance of Understanding Different Ways of Knowing

Scientifically literate citizens should have a reasonable grasp of the models generated through science and understand how these models are created. They should also be able to compare and contrast scientific models with other kinds of knowledge models, such as those identified in the preceding sections.

It is probably evident that scientific models and other kinds of knowledge models are equally important in human affairs. In fact, most issues are problems precisely because they must be decided by reconciling a number of competing demands encap-

sulated in these differing models. No significant issue is purely a scientific one.

Global climate change, for example, involves resolution of a complex of cultural, social, political, and economic conflicts; the science of its resolution is perhaps the least problematic issue at this point. Problems related to human population growth, energy resources, food production, and space exploration are also problems that would involve changes in social, cultural, economic, and political models to resolve.

In all these issues, the proper role of science is to provide the model(s) needed to understand the physical nature of the problem(s). Technologists may use this knowledge to develop technical solutions, politicians to provide political solutions, and so forth. The solutions to most of our most pressing problems require adjustments in many models that underlie our beliefs and actions.

This contextualization of science among other ways of knowing is important if we are to develop and retain a balanced perspective on what science can and cannot accomplish. We can develop contextualization skills by

- establishing that knowledge in any field is a culturally dependent set of diverse conceptual models;
- ensuring that students can delineate characteristics of scientific models from other kinds of knowledge models; and
- engaging students in analyzing the multiple relevant conceptual models underlying arguments on both sides of significant issues.

A Case Study: Mr. Thompson's Lesson

Mr. Thompson has introduced the idea of conceptual models to his tenth grade students. Now he wants to classify models by type so students will understand that scientific models are one among many other categories of conceptual models.

The Discussion

"John," Mr. Thompson calls on a student. "You're a pretty good soccer player. In fact, I hear that you're going to be on the varsity team next year."

John shrugs sheepishly and grins. "I guess," he replies.

"Well, let me ask you a question, then. When you're on the soccer field, how do you know what to do?"

"What do you mean?"

"Well, what actions do you take? Do you follow the ball? Watch another player?"

"Sure."

"Why don't you just stand still on the field?"

"That would be kind of stupid. It wouldn't work."

"Why not?"

"Well, when you stand still, you're not helping the team."

"So there are things you would do you think might work. There's a set of attitudes, expectations, and actions you might take to help your team score a goal?"

"Yeah, I guess so."

Mr. Thompson nods as if this were an important revelation, then continues. "In other words, when you're playing soccer, you work from a kind of model in your head. This is the way the world is and should be. How do you know if you're successful?"

John laughs. "I don't know. If I help the team score a goal."

"One more thing. Where did you get this model? Where did it come from?"

"I guess from lots of places. From the coach and the practice sessions. Experience mostly. Just knowing what to do and how to do it."

"Did you study it in books? *How to Win at Soccer*?"

John snorts scornfully. "No!"

"So this is a model that you built mostly from experience." Mr. Thompson now turns to the class. "We all have things we are successful at because we've learned how to operate in the world to reach our goals. We have ideas about how to do things, and these ideas are models about how things are, how they should be, and how we might behave to achieve certain things. Alan, what's the right thing to do—notice I said the right thing and not the only thing—if you find someone's notebook on the floor?"

Alan glances at his buddies. "I know the right thing to do would be to turn it in."

"What else could you do?"

"I could just throw it away."

"If someone saw you pick up the notebook, what would you do?"

"Turn it in."

"Why?"

"I could get in trouble if I threw it away, since somebody knew I had it."

Mr. Thompson nods. "So you have a scenario, a kind of model, already worked out in your head for what you might do and under what conditions, right?"

Alan nods.

"The models we carry around inside ourselves are what we call conceptual models. All of our models together make up the sum total of our worldview. Remember, we construct the world from our perceptions." Mr. Thompson stops for a moment and looks around the classroom. The expressions on the faces of some of his students tell him they're puzzled. "Look, let me explain this more completely. I'm going to expand my explanation because I see that some of you look a bit confused. What rule for teaching do you think I'm following right now? Joanne?"

Joanne hesitates. "I guess that if you see puzzled looks, then you should try to explain things in a different way."

"So the rule I'm following is part of my personal model of good teaching. I've built that model through experience, reading, and watching and listening to other teachers. Something is true in this model if what I do results in better student learning. It's false if it doesn't. I realize that there isn't any one way to be an effective teacher. Other teachers have different models. My conceptual model of teaching is made up of everything I know that influences how I behave as a teacher."

Jane raises her hand. "So everything we do is based on a conceptual model?"

"Things that you do consciously, at least, and some you do without thinking. Yes, because who we are and how we behave is all based on our mental models—what's up here in our heads. Now science is a systematic way we have developed to build these conceptual models in our heads. People used the processes of science well before we called it "science," but maybe not as systematically as we do today. There are rules we follow that are part of our model of what science is. What are scientific models for?"

The class is silent for a long moment. At last, Jamie raises her hand. "I think they're to find out what the world is about. You know, to find out what's true and what's not."

"So what does 'true' mean? Does anybody remember from our past discussions? Rod?"

"It just means everything's lined up right. Everything's in order."

"So what is truth in science?"

"I think it means that the way we explain things is all . . . lined up. Logical."

Claire chimes in. "Maybe truth in science is when what we *think* is happening and what we *see* happening line up with each other. There's no magic in it or anything like that."

Rod nods. "Yeah, you can be logical and reach funny conclusions. It's not just logic. You have to explain things based on what you can see."

"Or touch or hear or taste, et cetera," Nolan interjects.

Mr. Thompson nods, pleased.

"Then in science, the goal is to create models for what?"

"Just to explain things," John says.

"So we can explain the question of God, the universe, and everything using scientific models? Don't religious models explain things?"

"They do," Janet speaks up, "but they aren't based on just what we see. Science models explain the world we can sense. That's all. The same thing has to be able to happen over and over in a scientific model."

"So the assumption of science is that the universe operates in a predictable way and that under exactly the same conditions, the same thing will always happen? If that's the case, when we collect data from an experiment, are they always the same for everybody?"

"Well, no. But we aren't always doing things exactly the same way."

"Let's go back for a minute. How do you know if something is the truth?"

"Something's true if it explains things that we see. It doesn't have to explain everything there is to know."

"Okay. So how is a scientific model different from, say, a religious model?"

The students think about this for a long moment. Claire speaks up first. "A religious model wants to explain things about God. It doesn't depend just on what we see."

"If it doesn't depend on what you see, how do people decide what things to accept in your religious model?

"From the Bible or the Koran or somewhere like that. What they hear in church."

"Jeff?"

"Some religions think God speaks to you."

Sara asks, "Are we talking about religions, like, you know, Protestant, Catholic, and stuff? Or are we talking about religious models that individual people might have?"

"Good questions, Sara. And here again we have to distinguish between individual and shared group models."

"'Cause some people are religious but don't go to church. I think you can sort of sense God."

Mr. Thompson nods. "Now I don't want to get too deeply into this, because I don't want to say anything that might offend somebody. The important point to make is that religious models have different purposes and different rules that govern what goes into them than do scientific models. Can you really know something just by sensing it, Sara?"

Sara answers quietly. "I think so. I think there are different ways we know things. Science is one of them, but it's not the only way."

"All right," says Mr. Thompson. "Now it's time to put our discussion to work. I'd like to have us all analyze a few different kinds of conceptual models and see if we can figure out what the differences are among them. This activity might help us see how science fits in with all of the other ways we might model the world. For instance, Jack, you want to be a journalist. A journalist constructs a model of the world when he or she reports. How is that model different from a scientific model? Margie, when you study dentistry in a few years, you'll be learning a technological model. How is that different from a scientific one? I think this assignment will get us started answering these questions."

The Assignment

The assignment Mr. Thompson gives his class is the chart shown in Figure 7.1. In it, he has selected four types of conceptual models for analysis: literary, social, religious, and scientific. There are many others he might have chosen, such as political models (models of power); technical models (how to do things); mathematical models; emotional models; and parenting models, to name a few.

Mr. Thompson's purpose in this assignment is to engage students in an exploration of science as a way of knowing. An important part of this exploration is to distin-

Figure 7.1

Ways of Knowing: An Analysis

Type of model	What is the target of this kind of model?	Why is this kind of conceptual model important?	How is this model created?	What are some significant pieces of this kind of model?	How do we know if the model is true (fits its target)?

Figure 7.2

Completed Chart of "Ways of Knowing: An Analysis." Responses may vary.

Type of model	What is the target of this kind of model?	Why is this kind of conceptual model important?	How is this model created?	What are some significant pieces of this kind of model?	How do we know if the model is true (fits its target)?
Literary models (fiction)	A mental model created by the author.	These are stories we enjoy and from which we can learn and extend our experiences.	The author creates it using imagination and perhaps stories previously known.	Plot, character, setting.	The story feels plausible to us.
Social models (personal)	Behaviors resulting in effective social relationships with others.	Enhances our ability individually to survive and attain our goals.	Observation of behaviors of others; personal experiences; instruction by peers and authorities.	Social rules, customs, and norms; body language; verbal communications; dress rules; expectations of others.	We are able meet our goals with others.
Religious models (Protestant Christian)	Our relationship to God.	To save our souls when we die.	Direct instruction; admonitions of others; reading of scripture; revelations and intuition.	Analogies; rules and norms; articles of faith; stories; proverbs; parables.	Don't know since truth is evident only upon death.
Scientific models	Relationships in the natural world.	Allows us to understand and possibly control phenomena.	Through empirically validated research and experimentation.	Observations; data; inferences; predictions; hypothetical and theoretical models.	Model is repeatedly predictive and consistent with observations.

guish examples from nonexamples, for it is insufficient to provide learners with examples of a given concept without providing related nonexamples. For example, if we show a child a tree and say the word "tree" repeatedly, the child will eventually associate the word with the object. However, if we then show the child a bush, the child may well call it a tree. To know a tree, it is important to be able to distinguish it from a bush.

In this vein, students must be able to distinguish science from other fields of intellectual learning, such as theology, philosophy, medicine, economics, law, literature, and so forth. Students must keep in mind that science has become a model for the way to study things in many fields, such as economics or psychology. In other fields, however—which may include law, theology, art, or literature—science may not be the best method to use to build knowledge models. In fact, many of the personal conceptual models we create for ourselves, and use to guide our lives, are not scientific.

In Figure 7.2, Mr. Thompson has completed his own chart to use as a basis for class discussion.

After students have worked in small groups to complete the chart, Mr. Thompson brings the class together for discussion. He calls on each group to present its chart and explain the nature of group members' ideas. He hopes students will be able to expose disagreements among groups and eventually create a consensus. After group members have reviewed and revised their charts, Mr. Thompson asks them to summarize their understanding of the meaning of this assignment.

Conclusion

Readers may wonder how this chapter fits into a book entitled *Understanding Models in Earth and Space Science.* They may recognize that the chapter is not "science" in the sense of a course in biology, physics, or Earth science, but rather it is philosophy. These readers would be correct. Many teachers are reluctant to introduce their students to philosophy, and such study is seldom a part of standard science curricula. It is unfortunate that philosophy has come to be associated with sophism and argument for argument's sake. Although we do not think it is necessary to discuss philosophy by that name in our science classes, we do recognize that the essential ideas described in this chapter are philosophical in origin. We have included this chapter because we believe that, to understand science as intelligent consumers, we must recognize how it differs from other kinds of knowledge, and also how it is similar. The idea of knowledge as a structure of models says something about the nature of knowledge that we need to communicate, if we are to understand the limits of science and our own limits of knowing.

Appendix

References and Readings

American Association for the Advancement of Science. 1993. *Benchmarks for science literacy*. New York, NY: Oxford University Press.

Barber, B. 1961. Resistance by scientists to scientific discovery. *Science* 134: 596–602.

Bower, G. H. & Morrow, D. G. 1990. Mental models in narrative comprehension. *Science* 247: 44–48.

Bross, I. D. 1953. *Design for decision*. New York: The Macmillan Company.

Cartier, J., Rudolph, J. & Stewart, J. 2001. *The nature and structure of scientific models*. Working paper. The National Center for Improving Student Learning and Achievement in Mathematics and Science. Madison, Wisconsin.

Craik, K. 1943. *The nature of explanation*. Cambridge: Cambridge University Press.

Dreistadt, R. 1968. An analysis of the use of analogies and metaphors in science. *The Journal of Psychology* 68: 97–116.

Dreistadt, R. 1974. The psychology of creativity: How Einstein discovered the theory of relativity. *Psychology* 11(3): 15–25.

Duit, R. 1991. On the role of analogies and metaphors in learning science. *Science Education* 75(6): 649–-672.

Franco, C. & Colinvaux, D. 2000. Grasping mental models. In *Developing models in science education,* eds. J. K. Gilbert and C. J. Boulter, 93–118. Norwell, MA: Kluwer Academic Publishers.

Genter, D. 1979. *The structure of analogical models in science.* Cambridge, MA: Bolt, Beranek and Newman Inc.

Gilbert, J. K. 2000. Positioning models in science education. In *Developing models in science education,* eds. J. K. Gilbert and C. J. Boulter, 3–17. Norwell, MA: Kluwer Academic Publishers.

Gilbert, J. K., Boulter, C. J. & Rutherford, M. 2000. Explanations with models in science education. In *Developing models in science education,* eds. J. K. Gilbert and C. J. Boulter, 193–208. Norwell, MA: Kluwer Academic Publishers.

Gilbert, S. W. 1991. Model building and a definition of science. *Journal of Re-*

search in Science Teaching 28: 73–79.

Glynn, S. M. & Takahashi, T. 1998. Learning from analogy-enhanced science text. *Journal of Research in Science Teaching* 35: 1129–1149.

Grosslight, L., Unger, C., Jay, E. & Smith, C. 1991. Understanding models and their use in science: Conceptions of middle and high school students and experts. *Journal of Research in Science Teaching* 28(9): 799–822.

Halloun, I. 1998. Schematic concepts for schematic models of the real world. *Science Education* 82: 239–263.

Harrison, H. G. & Treagust, D. F. 1996. Secondary students mental models of atoms and molecules: Implications for teaching chemistry. *Science Education* 80: 509–534.

Johnson-Laird, P. N. 1983. *Mental models.* Cambridge, MA: Harvard University Press.

Lawler, R. W. 1996. Thinkable models. *Journal of Mathematical Behavior* 15: 241–259.

Leatherdale, W. H. 1974. *The role of analogy, model and metaphor in science.* Oxford: North Holland Publishing Company.

Lorenz, K. Z. 1974. Analogy as a source of knowledge. *Science* 185: 229–234.

Newton, D. P. 1996. Causal situations in science: A model for supporting understanding. *Learning and Instruction* 6(3): 201–217.

Mathews, M. R. 1994. *Science teaching: The role of history and philosophy of science.* New York: Routledge.

Meadows, D. H. and others. 1972. *Limits to growth; a report for the Club of Rome's project on the predicament of mankind.* New York: Universe Books.

Morris, W. T. 1970. On the art of modeling. In *The Process of Model-Building in the Behavioral Sciences*, ed. R. M. Stogdill. New York: W. W. Norton & Co.

National Research Council. 1996. *National Science Education Standards.* Washington, DC: National Academy Press.

Noh, T. & Scharmann, L. C. 1997. Instructional influence of a molecular level pictorial presentation on matter on students' conceptualizations and problem-solving ability. *Journal of Research in Science Teaching* 34(2): 199–217.

Ost, D. 1987. Models, modeling and the teaching of science and mathematics. *School Science and Mathematics* 87(5): 363–370.

Pennen, D. E., Giles, N. D., Leherer, R. & Schauble, L. 1997. Building functional models: Designing an elbow. *Journal of Research in Science Teaching* 34(2): 125–143.

Raghavan, K. & Glaser, R. 1995. Model-based analysis and reasoning in science: the MARS curriculum. *Science Education* 79(1): 37–61.

Ritchie, S. M., Tobin, K. & Hook, K. S. 1997. Teaching referents and the warrants used to test the viability of students' mental models: Is there a link? *Journal of Research in Science Teaching* 34(3): 223–238.

Rudolph, J. L. & Stewart, J. 1998. Evolution and the nature of science: On the historical discord and its implications for education. *Journal of Research in Science Teaching* 35(10): 1069–1089.

Shuell, T. J. 1990. Phases of meaningful learning. *Review of Educational Research* 60(4): 531–547.

Solomon, J. 1995. Higher level understanding of the nature of science. *School Science Review* 76(276): 15–22.

Stevens, A. & Collins, A. 1980. Multiple conceptual models of a complex system. In *Aptitude, Learning and Instruction,* (Volume 2), eds. R. Snow, P.A. Federico & W. Montague. Hillside NJ: Lawrence Erlbaum Associates Inc.

Van Driel, J. H. 1998. *Teachers' knowledge about the nature of models and modeling in science.* Paper presented at the annual meeting of the National Association for Research in Science Teaching, San Diego, CA.

Vedantum, S. 2002, October 14. When sound is red: Making sense of mixed sensations. *The Washington Post*, A12.

Weller, C. M. 1970. The role of analogy in teaching science. *Journal of Research in Science Teaching* 7: 113–119.

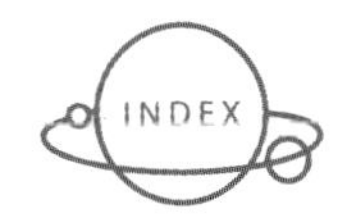

Index

Page numbers in **boldface** type refer to figures and tables.

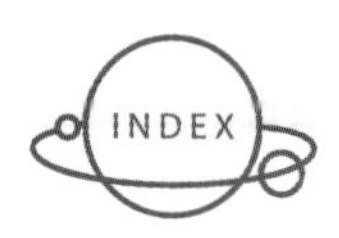

D

E

F

G

H

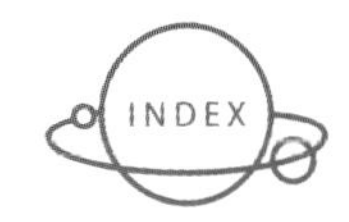

I

K

L

M

N

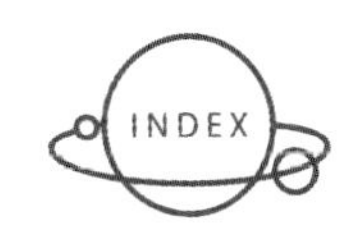
INDEX

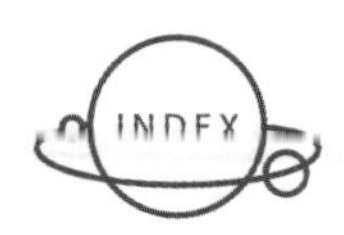

T

U

V